Confabulations, Lies, And Conspiracy

Carsten R. Jorgensen

DEDICATION

This book is dedicated to my daughter, Heather, who introduced me to these topics and makes me stop to think about all of the unusual things that go on in our world.

CONTENTS

PART ONE – CONFABULATION

1 Confabulation   3

PART TWO – LIES

2 Lies   7

3 Hypnosis   11

PART THREE – THE MANDELA EFFECT

4 Who Was Nelson Mandela?   15

5 The Mandela Effect   25

6 Explanations For The Mandela Effect   43

PART FOUR – ALTERNATE UNIVERSES

7 The Alternate Universes   49

8 Adventures In An Alternate Universe   52

PART FIVE – CONSPIRACY

9 Conspiracy   65

10 Chidren's Television   67

11 Area 51 And UFOs   70

12 NASA Faked The Moon Landing   81

13 Reptilian Humanoids   91

14 Secret Societies   93

15 The JFK Assassination   98

16 The Plot To Kill Fidel Castro   115

17 The CIA And AIDS   117

18 The 9/11 Cover Up   118

19 Paul McCartney Imposter   125

20 Jesus And Mary Magdalene   127

21 Holocaust Revisionism   129

22 The Sandy Hook Elementary School Massacre   131

23 Some Final Thoughts   135

    References   137

PART ONE :

CONFABULATION

1. CONFABULATION

As we age, some of our memories slip away. Society tells us that we should be resigned to getting older. Studies show that the mind stops making new neural connections after age 40 and relies on the existing ones. We begin to experience senior moments such as forgetting where we left our keys or losing the word that was on the tip of the tongue but is now out of reach.

Seniors are not the only ones plagued with memory problems. Some people have false memories called confabulations. Confabulation is a term in cognitive psychology defined as a recollection of something that never happened. This can be very minor such as incorrectly remembering an item on a list, or it could be more major, like fabricating an entire vivid memory. It is intuitively obvious that memory is fallible. It is not like a tape recorder. It is a process that reconstructs past experiences. Thus, it is highly susceptible to errors.

Paranormal Consultant, Fiona Broome, draws comparisons between existence and the holodeck of the USS Enterprise from Star Trek. The holodeck on Star Trek is a virtual reality system, which creates a recreational experience. By Broome's explanation, memory errors are software glitches.

Confabulation occurs when the brain is attempting to fill in the blanks for incomplete memories. The speaker may mix and match similar experiences and information in order to complete the story in his or her mind, complete with details and emotional responses. They become certain that the tale is true. This kind of behavior happens in people suffering from neurological issues, such as brain damage or Alzheimer's, but healthy individuals confabulate, too.

So, this raises some questions. How much of what we remember is real and how much has been confabulated in our heads? When someone tells what you consider to be a lie, are they always doing so consciously, or do they really believe what they are saying is true sometimes? We can forgive a person with obvious mental problems, such as Alzheimer's, but what of those who are not diagnosed with such diseases?

I can recall clearly a time when a friend was recounting a teenage event that happened to both of us. However, when she told the story, many of the facts were different from how I remembered them. I did not interrupt her and correct her because I knew that people remember things differently. I considered that there wasn't any harm in her facts being different from what actually happened and what difference did it make anyway. To be clear, if the matter had been of importance, I would have spoken out and corrected her. But as far as I could tell, that was the way she remembered it, and I was fine with that. Each person perceives the world differently and experiences things differently. It is like we are all in our own individual movie; but no one is in the same movie.

People who confabulate are usually very confident about what they remember, despite evidence to the contrary. So, what if what they remember is actually true? As I grew older and learned about things like the Mandela Effect (which we will come to a little later), I began to wonder if these differences in memories were confabulations, lies, or something more.

PART TWO :

LIES

2. LIES

I have been told that everybody tells lies. Most lies are called little white lies. These lies are told to keep a friend from having hurt feelings. Most people consider these lies harmless. Lies told in order to make personal gains are usually condemned.

Most lies are told to avoid punishment. One of the most well-known, stereotypical lie might be the public school student who said, "The dog ate my homework".

Avoiding humiliation or embarrassment in conjunction with punishment is a common reason people tell deliberate lies. With deliberate lies, there is a threat of facing consequences if the lie is discovered. These consequences can be small, such as losing face, but they can also be larger, such as a loss of a relationship, freedom, money, career, or even a loss of life itself.

In 2015, Julia Shaw, and colleague, Stephen Porter, wrote an article on constructing rich false memories. They explained that suggested or imagined elements of a memory with reference to what something could have been like, could translate into what an event would have been like, which then, can eventually turn into what was perceived to actually have been like. After three interrogation sessions, 76% of test subjects recalled a false emotional event and 70% recalled a fictional crime. Shaw and Porter did not find any distinguishing personality traits of people who made false statements and did not find a way to single out individuals who are the most inclined to do so. It is possible that false memories may be recalled in the brain in a similar fashion that true memories are retrieved, so it can be difficult to tell the difference between the two.

Thus, when police interview a suspect, they might (knowingly or unknowingly) give the suspect information which is related to the crime being investigated. This could create a false memory in the suspect. According to Shaw and Porter, in 70% of these interviews, the suspect would develop a false memory of the crime.

Information that you learn after an event can change your memory of the event. This includes even subtle information and can make a witness statement unreliable. In this way, an interrogator may be able to affect a witness's memory of an event.

"Priming" refers to circumstances or elements that happen before an event which affects your memory of an event. Studies have shown that suggestions made by an interrogator can affect a participants' recall. The police can use the suspect's statements from the interrogation. Choice of words in interrogation may affect the witness.

Priming is also called suggestibility and presupposition. For instance, if you were to ask how short a person is instead of how tall a person is, this would influence a subject to think that the person in question was short. Even just changing a single word can make a profound difference in how a situation is recalled. If you ask, "Did you see the black car?" instead of "Did you see a black car?" you are making a suggestion in the person's mind that it is a fact that the car was black. This affects how people respond. In this way, suggestions and misleading statements can affect your existing memories.

An example of how just one word making all the difference in influencing memory recall can be seen in a study that was done involving a car accident. One group of witnesses were asked "How fast were the cars going when they hit one another?". The other group of witnesses were asked "How fast were the cars going when they smashed into one another?". The witnesses who were asked the question using the word "smashed" all recounted seeing glass on the ground. The witnesses who were asked the question using the word "hit" all perceived the cars to be travelling slower and none of them recounted seeing glass on the ground.

Criminal lawyer, Natasha Calvinho, has the following advice on her Facebook page:

HOW TO TALK TO POLICE:

1) DON'T.

In the United States, thousands of people plead guilty every year, not because they actually are guilty, but because they know that there is not much chance of a jury believing their testimony over that of a police officer. As one lawyer stated, "Everyone knows you have to be crazy to accuse the police of lying". Indeed, this happens so much, that there are documentaries and television shows outlining numerous examples of this. Those are just the ones that are documented on film. There are thousands of innocent people's stories that go untold. A lucky few are able to prove their innocence but by that time they have lost years of their lives due to these lies. And can we repay these victims of the lies once they are proven innocent? We can't. Nothing can be done to give them back those years. This is why it is so important to have trustworthy police officers working to serve and protect the people.

But police officers are no more trustworthy than anyone else. They have an incentive to lie. In this day of mass incarceration, the police shouldn't be trusted any more than any other witness; perhaps less so.

Many honest law enforcement officers are aware of what some of their not so honest counterparts are doing. A former San Francisco Police commissioner, Peter Keane, wrote an article in The San Francisco Chronicle decrying a police culture that treats lying as the norm. "Police officer perjury in court to justify illegal dope searches is commonplace. One of the dirty little, not-so-secret, secrets of the criminal justice system is undercover narcotics officers intentionally lying under oath. It is a perversion of the American justice system that strikes directly at the rule of law. Yet, it is the routine way of doing business in courtrooms everywhere in America."

3. HYPNOSIS

"Hypnosis involves a person's ability to set aside critical judgment without relinquishing it completely, and to engage in make-believe and fantasy". (Gill & Brenman, 1959; E. R. Hilgard, 1977).

For some people, this make-believe may be so lifelike and profound that they may have trouble discerning it from what is real.

The experience of hypnosis has very little to do with the hypnotist. It has mainly to do with the ability of the person hypnotized. The ability to conduct a hypnotic induction is learned easily and rapidly by an individual who has moderate interpersonal skills and can establish a relationship of trust and the appearance of competence.

Hypnosis has three main effects upon memory.
1) It increases productivity; but most of the new knowledge is inaccurate.
2) Confidence is increased for both correct and incorrect "novel" remembrances.
3) The accumulation of confidence is found at all levels of hypnotizability.

Since there is a tendency to confuse fantasy as fact, the result of hypnosis can be a confabulation. More studies are needed to see to what extent that hypnosis has resulted in a confabulation, if any. There is a possibility that these studies will find this to be true, but it could also be a lie. There are four possibilities – truth, lie, confabulation, or pseudo-memory. Each of these possibilities requires meticulous assessment. Many issues regarding hypnosis and memory are argumentative among those studying in this field.

Twenty-five American State Supreme Courts out of thirty, that have heard cases that had a hypnosis component, have placed per se (that is, automatic) exclusions on hypnotically elicited testimony (State v. McClure, 1993).

PART THREE :

THE MANDELA EFFECT

4. WHO WAS NELSON MANDELA?

Nelson Mandela in Johannesburg, Gauteng, on 13 May 2008

In the village of Mvezo in the eastern Cape, South Africa, a woman named Nonqaphi Nosekene, gave birth to a boy named Rohlihlala, on 18 July 1918. His father, Nkosi Mphakanyiswa Gadla Mandela, was the predominant counselor to Jongintaba Dalindyebo, the Acting King of the Thembu people.

Since his parents were Christians, they sent Rohlihlala to a

Christian primary school in Qunu. There, his teacher, Miss Mdingane gave him his Christian name, "Nelson". It was custom to give all schoolchildren "Christian names".

In 1930, when Rohlihlala was 12 years old, his father died. As per his father's wishes, his mother entrusted Nelson's care to the Regent of Thembuland Jongintaba, David Mtivava, his wife, NoEngland Mtivava in Mqhekezweni. He shared a room with Justice, the Regent's son. In a nearby school, he studied things like Geography, English, and History . It was here, he would listen attentively to the stories of his country; and it was here that he fell in love with the history of Africa.

After two years at Clarkebury Methodist High School, where he had studied to acquire the skills of a counselor, Nelson then registered at a Healdtown, a Wesleyan secondary school in Fort Beaufort.

Nelson then attended the University College of Fort Hare and began his studies for a Bachelor of Arts degree. However, he did not complete his degree there because he was expelled for joining a student protest over the process for electing prefects.

The Regent was furious and said that if he did not return to Fort Hare, he would arrange for wives for him and Justice. Nelson and Justice ran away to Johannesburg where Nelson worked as a mine security officer in 1941.

Nelson then attended the University of South Africa, where he finished his B.A. He went back to Fort Hare in 1943 for his graduation. After graduation, he began studying at the University of the Witwatersrand for an LLB degree. Nelson conceded that he was a poor student and decided to leave the University of the Witwatersrand in 1952 without graduating.

Nelson became good friends with Walter Sisulu, who was an anti-apartheid activist. Nelson was the best man at Walter's wedding. In 1944, Nelson married Walter's cousin, a nurse, named Evelyn Mase. They had two sons and two daughters. The first of the daughters died in infancy. He and Evelyn were divorced in 1958.

Nelson Mandela in 1937

The year 1952 saw many changes for Nelson. In that year he was selected to be the National Volunteer-in-Chief of the Defiance Campaign. Maulvi Cachalia was chosen as his deputy. The Defiance Campaign was a joint program between the ANC and the South African Indian Congress and consisted of acts of defiance against six laws that were unjust. The campaign was launched June 26, 1952, on the anniversary of the National Day of Protest. The campaign included boycotts of buses in South Africa, socializing in places that were considered "whites only", and the burning of pass books. Nelson and 19 others were arrested and charged under the Suppression of Communism Act for their part in the campaign.

Nelson had achieved a two-year diploma in law on top of his BA which allowed him to practice law. In August 1952, Nelson and Oliver Tambo started South Africa's first black law firm, Mandela & Tambo.

In December 1952, the sentencing for his part in the Defiance Campaign took place. He was sentenced to nine months of hard labour, suspended for two years. Additionally, he was banned for six months under the Riotous Assemblies Act. That was only the first time he was banned. As soon as that ban was finished, he immediately threw himself into another campaign. This time the campaign was against the forced removal from Sophiatown and Western Areas. This resulted in his second ban. As a restricted person, he was only permitted to watch in secret as the Freedom Charter was adopted in Kliptown on 26 June 1955.

On 5 December 1956, in the early hours of the morning, there was a country wide police sweep. Nelson Mandela's house was raided, and he was arrested. There were 156 men and women of all races charged with high treason, and they were put on a marathon trial, called the 1956 Treason Trial. They were all granted bail and the court proceedings were set for January 1957. During the trial, Mandela married South Africa's first black, female, social worker, Winnie Madikizela. The trial lasted for 5 years and finally ended 29 March 1961, when the last 28 accused, including Mandela, were acquitted.

On 21 March 1960, police killed 69 unarmed people in a protest in Sharpeville against the pass laws (a form of internal passport system designed to segregate the population. Pass laws severely limit the movements of people by requiring them to carry pass books when outside their homelands or designated areas). This event led to South Africa's first declared state of emergency and the banning of the ANC and the Pan Africanist Congress (PAC) on the 8th of April. Mandela and his colleagues in the Treason Trial were among thousands detained during the state of emergency.

In June 1961 Nelson was asked to lead an armed conflict. He helped to establish Umkhonto weSizwe (Spear of the Nation). On 16 December 1961 the battle began with a series of explosions.

On 11 January 1962, using the adopted name David Motsamayi, the ANC secretly sent Mandela as a delegate to Ethiopia to the Pan-African Freedom Movement for East, Central and Southern Africa (PAFMECSA). He travelled around Africa and visited England where he met anti-apartheid activists, reporters, and prominent politicians to gain support for the armed struggle. He received military training in Morocco and Ethiopia and received funds from Liberian President, William Tubman, and Guinean President, Ahmed Sékou Touré. Nelson returned to South Africa in July 1962 where he briefed ANC President Chief, Albert Luthuli about his trip. Upon returning from this briefing, he was arrested in a police roadblock outside Howick on August 5.

He was charged with inciting workers' strikes and leaving the country without a permit. He was pronounced guilty and sentenced to five years in prison. He had been jailed in Johannesburg's Marshall Square prison but later was moved to the Pretoria Local Prison where his wife, Winnie, was able to visit him.

Within a month, police raided Liliesleaf, a secret hideout in Rivonia, Johannesburg, used by ANC and Communist Party activists, and uncovered documents that outlined the activities and mentioned Mandela. Mandela and ten of his comrades were charged with four counts of sabotage and conspiracy to violently overthrow the government. This became known as the Rivonia Trial and it began on 9 October 1963.

Mandela used the trial to bring their cause to the world. His words to the court at the end of his famous three-hour "Speech from the Dock" (also known as his "I Am Prepared To Die" speech) on 20 April 1964 became immortalized as it was published in newspapers all over the world:

"I have fought against white domination, and I have fought against black domination. I have cherished the ideal of a democratic and free society in which all persons live together in harmony and with equal opportunities. It is an ideal which I hope to live for and to achieve. But if needs be, it is an ideal for which I am prepared to die. " - *Speech from the Dock quote by Nelson Mandela on 20 April 1964*

On 11 June 1964, Mandela and seven other accused men (Walter Sisulu, Ahmed Kathrada, Govan Mbeki, Raymond Mhlaba, Denis Goldberg, Elias Motsoaledi and Andrew Mlangeni), were convicted. The prosecution had called for the death penalty but instead they were sentenced to life imprisonment. Mandela and all but one of his co-accused were incarcerated in Robben Island Prison. Goldberg was sent to Pretoria Prison because he was white.

A room in Robben Island Prison

They were kept isolated from non-political prisoners in their own section. They would break rock into gravel each day. The glare from the sun off the lime stone eventually did damage to Mandela's eyesight. Mandela's mother died in 1968 and his eldest son, Thembi, died in 1969. He was not allowed to attend either of their funerals.

On 31 March 1982 Mandela was transferred to Pollsmoor Prison in Cape Town along senior ANC leaders Sisulu, Mhlaba, and Mlangeni. Kathrada joined them in October. Conditions at Pollsmoor were better than Robben Island.

In November 1985, Mandela underwent prostate surgery. After his return to prison, Mandela was held alone. Justice Minister Kobie Coetsee secretly visited him in the prison. Coetsee organized negotiations between Mandela and some government figures. They were willing to release the political prisoners on the condition that they permanently break links with the Communist Party, renounce violence, and not insist on majority rule. This was not agreeable to Mandela.

Nelson Mandela In His Prison Cell

On 12 August 1988, Mandela contracted tuberculosis which was made worse by the dampness of his cell. By this time he was over 70 years old. In December, he was transferred to a warders house at Victor Verster Prison.

In 1989, Botha suffered a stroke and was replaced by the new president, F.W. de Klerk. He believed that the aparthied was unsustainable and as a result he released some of the ANC prisoners. By November 1989, de Klerk began the legalization of the ANC and all formerly banned political parties. On Sunday, 11 February 1990, Mandela left Victor Verster Prison on an unconditional release He held hands with Winnie in front of amassed crowds and the press and the event was broadcast live around the world.

After his release, Mandela occupied his time in trying to end white minority rule. In 1991, Oliver Tambo fell ill and Mandela was elected ANC President to replace him. In 1993, he and President F.W. de Klerk jointly won the Nobel Peace Prize.

On 27 April 1994, for the first time in his life, Mandela was able to vote. Subsequently, on 10 May 1994, he was inaugurated as South Africa's first democratically elected President.

Nelson and Winnie divorced in March 1996. He married his third wife, Graça Machel, on his 80th birthday in 1998. In 1999, Mandela stepped down after only one term as President. However, he continued to work with the Nelson Mandela Children's Fund he had set up in 1995. He also created the Nelson Mandela Foundation and The Mandela Rhodes Foundation.

Nelson Mandela died at his home in Johannesburg on 5 December 2013.

When Mandela died, many people watched his funeral procession on TV. Some of these people remembered that he died in the 1980s. They also remembered his wife giving a eulogy. Although many people remember that Nelson Mandela died in the 1980s, he did not die at that time.

President Nelson Mandela

5. THE MANDELA EFFECT

You may be wondering " What is the Mandela Effect?" The mandela effect is a term coined by the author and paranormal researcher, Fiona Broome. While at a conference, the topic of Nelson Mandela's death came up. Fiona mentioned how she had thought that former South African president, Nelson Mandela, had died in a South African prison in the 1980s. She told the people at the conference that she had been surprised to find out that she was wrong and that Nelson Mandela had actually died on 5 December 2013. However, she was more shocked to find that so many other people also remembered the event of Nelson's death taking place in the 80s, and that she was not alone in this memory. Many people not only believed that Nelson Mandela had died in the 1980s, but they remembered it in great detail. Many say that they saw his funeral procession on TV at that time. They remembered news coverage of the event and they also remembered a speech given by his widow.

After this discovery, Fiona wrote a book about it and started a website to discuss other instances like it. She called the phenomenon "the Mandela Effect". Her explanation is that it is a malfunction or glitch in our memory. However, many people believe this phenomenon is something more, which we will come to later.

Through Boome's website other group false memories emerged. As time went on, Fiona was collecting more and more examples of this phenomenon.

Henry VIII Eating A Turkey Leg

People had a memory of a painting showing Henry VIII eating a turkey leg. No such painting exists. However, there are cartoons of him eating a turkey leg.

Someone did post a picture on the internet of Henry VIII with a turkey leg, but it is thought to be photoshopped.

Perhaps the "memory" of King Henry VIII holding a turkey leg comes from the 1920 painting of him holding a brown glove (as seen on the previous page). The glove could easily be mistaken for a turkey leg.

"Luke, I Am Your Father"

In "Star Wars Episode V, The Empire Strikes Back", most people remember Darth Vader saying, "Luke, I am your father". However, the actual line was, "No, I am your father".

Yesterday, I was talking with a Star Wars fan who had just watched Star Wars Episode V, The Empire Strikes Back. According to him the line was, "I am your father".

Interestingly, during an interview in a clip from "Star Wars Empire of Dreams" (a documentary of the original Star Wars Movie Trilogy), James Earl Jones, the voice of Darth Vader, even recalls saying "Luke, I am your father". This can be viewed on Youtube. (https://www.youtube.com/watch?v=GQ1mmkKb_BQ)

As well, a Hallmark Christmas ornaments were sold in 2016 that had the recorded voice of James Earl Jones saying "Luke, I am your father". This can also be viewed on Youtube. (https://www.youtube.com/watch?v=dEnRV9P74XE)

Mirror, Mirror, On The Wall

In the Walt Disney movie "Snow White and the Seven Dwarfs" the wicked Queen addresses a mirror and says "Mirror, mirror on the wall". That is what people remember. However, she actually says "Magic mirror on the wall".

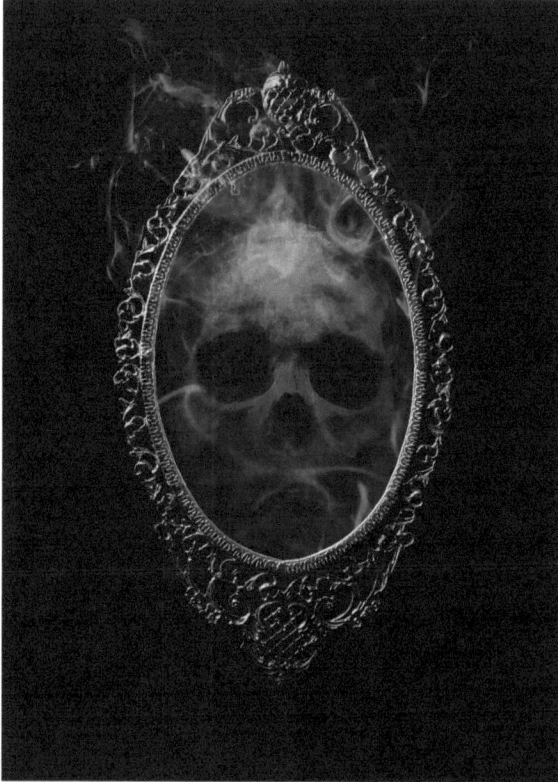

"Magic mirror on the wall"

"Oscar Meyer" Isn't Spelled That Way

Perhaps you remember as a kid singing the little jingle from the hot dog commercial. The lyrics went like this: ""My bologna has a first name, it's O-S-C-A-R / My bologna has a second name, it's M-E-Y-E-R!". You might be surprised to find out that it is actually spelled M-A-Y-E-R, with an "a". Which version of the spelling is the real bologna here?

The Show Isn't Called "Sex In The City"

It's **Sex and the City**, but many people insist they remember it being "in the" at some point. Some people have even posted pictures of old memorabilia they have that supports their false memory.

"We Are The Champions" By Queen Ends Differently

Many of those who are familiar with the song remember the final lyrics being "No time for losers, 'cause we are the champions...**of the world!**" There is no "of the world!" The song just ends, and it's driving people crazy because they feel 100% sure that they've heard otherwise in the past.

The Monopoly Man's Monocle

The game of Monopoly has a cartoon character, Rich Uncle Pennybags, that is featured on various components of the game. Many people remember him as having a monocle, but he doesn't.

It is possible that they are just confusing him with Mr. Peanut, the Planters peanut mascot. Both of them wear a top hat and carry a cane. However, there are a number of people who vividly remember the Monopoly man having a monocle and are taken aback when they find out otherwise.

Mr. Peanut Balloon Lands in Children's Park
InOttawa.ca / CC BY (https://creativecommons.org/licenses/by/2.0)

The Spelling Of Berenstain Bears

This is one of the more popular Mandela effects that is debated, in which some people seem to recall the book series/cartoon about a family of bears being known as **The Berenstein Bears**. However, if you look now, they're actually called **The Berenstain Bears.** Many folks insist they remember it being spelled with an "e," and one Reddit poster even found an old VHS tape of the cartoon, and the label shows "Berenstein."

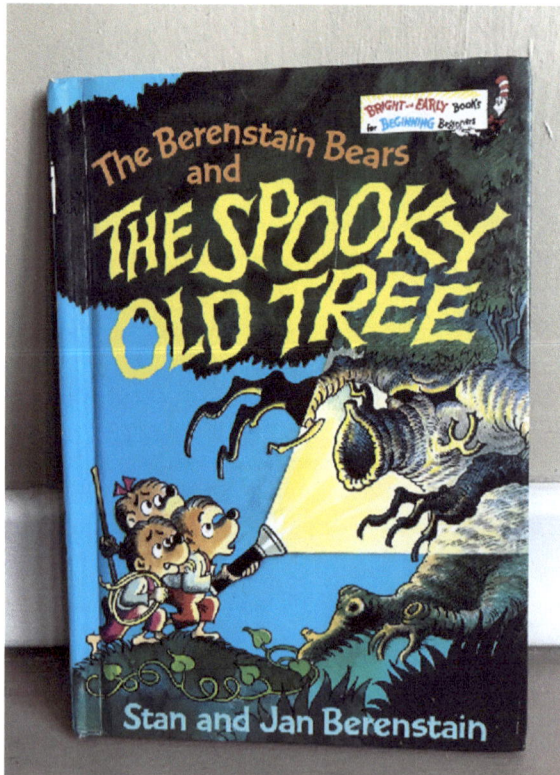

Picture of my daughter's Berenstain Bear book

The Berenstein Bears Learn About Strangers VHS

But perhaps people are remembering both because both spellings were actually in circulation at the same time. This could certainly explain the confusion in this instance.

Two different spellings on VHS tapes

Curious George Never Had A Tail

Another childhood favourite was Curious George. You remember how he would use his tail to swing from the trees, right? Well, if you look up pictures of Curious George today, you'll see that he doesn't have a tail.

Old World monkeys, except the Barbary macaque, also have tails. Apes (gibbons, siamang gibbons, gorillas, chimps, and orangutans) lack tails, as do humans. This is an important distinction because, as depicted, Curious George with no tail, suggests he is an ape or possibly a Barbary macaque. (Sep 30, 2016) But people remember him with a tail. Was his tail amputated or did he switch species?

The Tip Of Pikachu's Tail Isn't Black

People remember there being a black mark on Pikachu's tail, but if you take a look at Pikachu now, you'll see nothing there. How so many people can remember an aspect of this character's appearance that doesn't actually exist, the world may never know.

Pikachu as seen nowadays

Pikachu as remembered by many
(the tail in this image is photoshopped)

Mister Rogers' Theme Song Opening Line Is Different From What People Remember

You may recall the children's show, Mister Rogers' Neighborhood and the catchy little theme song that he sang at the beginning. But do you remember the words to that song? Perhaps you sing "It's a beautiful day in the neighborhood." That's not what it was, though. Instead, he sings, "It's a beautiful day in **this** neighborhood." For many, the difference between just that one word makes the song not feel quite right. There is a feeling that somehow this song has changed.

Life Isn't Like A Box Of Chocolates

Another instance of a single word being changed in a famous line is from the movie Forrest Gump. Many people are confident that his mama always said, "Life is like a box of chocolates." As it turns out, the line is actually, "Life **was** like a box of chocolates."

Hannibal Lecter Never Said "Hello, Clarice."

In the movie *The Silence of the Lambs*, when Clarice first meets Hannibal Lecter, many remember him saying "Hello, Clarice." It is one of the most famous lines from that movie. However, that never happened. He simply says, "Good morning." That's it. How is a film's most well-known line nonexistent? Nobody knows, and it's eating away at people.

Mona Lisa's Smile

Mona Lisa's smile is the most famous thing about her. You would think that everyone would remember it the same. However, a lot of folks are insistent that her smile has changed. They remember her having a straight face, but now they feel that she is smirking.

Mona Lisa by Leonardo Da Vinci

There are many reasons why Mona Lisa's smile may have changed. There have been more than a few conspiracies that the Mona Lisa had been stolen and replaced with a fake and then covered up.

It is also believed by some that a second painting of the Mona Lisa was made by Leonardo Da Vinci which preceded the one that now hangs in the Louvre. This claim was made by the

Mona Lisa Foundation in 2012. The painting hung in the home of an eccentric art dealer, Henry Pulitzer, for many years until his death. The ownership of the painting then went to Mona Lisa Inc in Anguilla. Currently, this second painting is the subject of an ongoing lawsuit by Andrew and Karen Gilbert, from London, who say they own a 25% share in the painting.

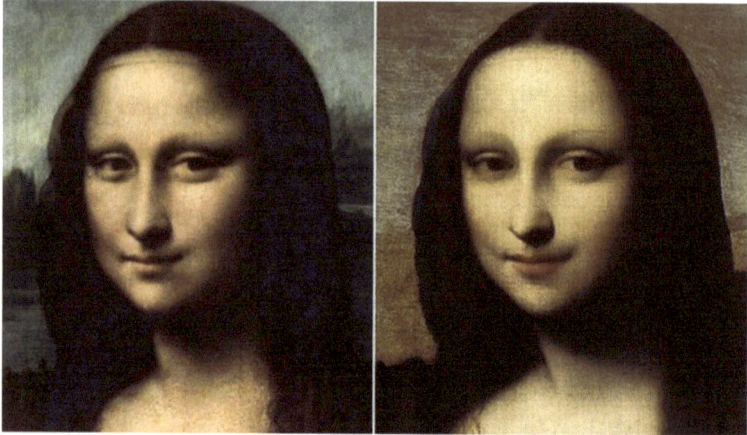

Leonardo Da Vinci may have painted two Mona Lisas

The problem is that the copyright for the Mona Lisa is in the public domain and so, anyone can take her image and change it as they see fit. They can then post it on the internet, use it in books, posters, T-shirts, or any other type of media. Nipissing University in North Bay Ontario once manipulated her image in the 80s as promotional material. They did this with an image of Albert Einstein as well. They had the images dressed in Nipissing University T-shirts, and they were used on posters and other promotional items. Since there are so many versions of the Mona Lisa out there now, who is to say which ones are true memories of the real one and which ones are memories of an altered version they have seen somewhere else.

Carmen Sandiego

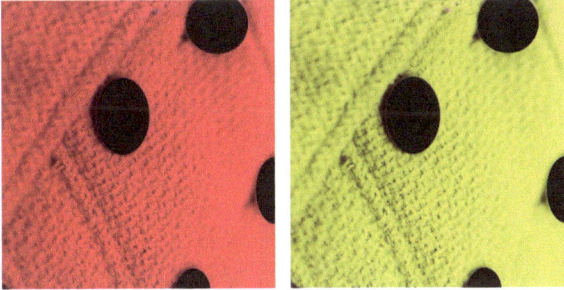

Carmen Sandiego is a fictional character from a children's show. She is always seen travelling around the world wearing her signature red trench coat and wide brimmed hat. The double-breasted coat is an unmistakable bold red. However, there are people who remember her wearing a yellow version at some point in time.

Chartreuse

Would the real chartreuse please stand up! Do you remember chartreuse as pink or green? Well, if you remember it as a yellow-green colour then you would be correct. Chartreuse gets its name from a french liqueur that was introduced in 1764.

The Crash Of United Flight 93 During The Sept. 11 Terrorist Attack

United Flight 93 was a passenger plane that was hijacked by four al-Qaeda terrorists as part of the September 11, 2001 attacks. In an experiment, about 30 percent of people affirm that they had viewed video footage of the crash of United Flight 93. Interestingly, no such footage exists. When asked if they remembered having seen the footage, 20 percent of people with highly superior autobiographical memory (those with an uncanny ability to remember dates and events) reported that they had indeed viewed the nonexistent video.

Tiananmen Square

Tank Man is the nickname given to an unknown Chinese protester in Tiananmen Square on June 5, 1989. What makes him stand out from all the other protesters is that after weeks of demonstrations in Beijing, he was captured on video, shopping bags in hand, stopping and standing by himself in front of a line of approaching tanks. The result was that the tanks stopped for a brief time. On this fact everyone agrees.

What happens next is what is debated by many. Some remember that the man was then run over by the tanks. Others remember the tanks getting dangerously close to him but then the man made a hasty retreat. What actually happened was the still-unidentified jump up on the tank, knocked on the lid and had a conversation with the soldier inside. He then jumped back down and resumed his place in front of the tank until a handful of protesters came to his side and ushered him away.

My Personal Experiences With The Mandela Effect

My daughter, Dana, asked me to proofread her recipe book. I took the book and opened it. On a page near the back of the book, near the bottom I saw a mistake. She had written the word "ON" twice so that it said "ON ON. " I took my pen and circled the second "ON.".I usually write the page number and my observation on a small note paper, but I did not have my note paper handy. So, after getting my note paper, I started at the beginning of the book to proofread for a second time. I never saw that page with ON ON again. It is not there.

Many years ago, a friend of mine hand carved a wooden viking dragon head for me. It was a one of a kind gift; the only one in existence. However, one day my daughter, Dana, who was about 8 years old, asked where the green dragon head was. She explained that she remembered a hunter green dragon head, identical to the red one and that it sat on the wall next to the red one, facing away from each other. Of course, I told her that a green one never existed. I was sure that she was just confused, but then her younger sister, Heather, who was 6 years old said that she remembered the green dragon too and had also wondered where it had gone.

Flip Flops

In some instances the Mandela Efeect does what has been termed as a "flip flop". This is where you remember something, such as a spelling of a product, find out that it has changed in spelling and then later on down the road discover that it has changed back to it's original form of spelling.

When something flip flops in the Mandela Effect, most often all former examples of the change disappears so that it can not be proven. However, sometimes not all proof of change gets erased and this is known as Mandela Effect residue.

The Mandela Effect seems to be not quite universal. To some the monopoly man somewhere is wearing a monocle and Pikachu somewhere has a black spot on his tail. The Berenstain Bears somewhere are the Berenstein Bears. As some would say, it all depends on what timeline you are in. Some people are really experiencing multiple realities.

6. EXPLANATIONS FOR THE MANDELA EFFECT

One explanation for the Mandela effect is a confabulation, which usually happens to a single individual. But this does not make for a good explanation when it happens to a multitude of people.

There are several theories which explain the effect. One explanation is suggestibility which is the tendency to believe what others suggest being true. This is why lawyers are prohibited from asking witnesses leading questions that suggest a specific answer. It has been suggested that information appearing on the internet may cause such suggestibility.

Other explanations fall within the realm of pseudoscience. One explanation involves time travelers from the future who have gone back and changed some things in our past. They kept Nelson Mandela alive while he was in prison. They got involved in some of our movies and a few things in them were changed. They got involved in the manufacturing of the game "Monopoly", etc.

The most recent explanation I have seen suggested is that all those people who saw Mandela's funeral procession in the 1980s somehow slipped into the future to 2013 and saw Mandela's funeral procession on TV. Then, shortly after, they slid back to their own time in the 1980s.

Another explanation is the multiverse. The concept of parallel realities or alternate universes. This has people going to an alternate reality where events have happened a little differently than in their original reality.

The concept of alternate realities is a relatively "new" theory, called the "many interacting worlds" hypothesis (MIW), and the idea is just as profound as it sounds. The theory suggests not only that parallel worlds exist, but that they interact with our world on the quantum level and are thus detectable. Though still speculative, the theory may help to finally explain some of the bizarre consequences inherent in quantum mechanics. This has also been used to explain the Mandela effect.

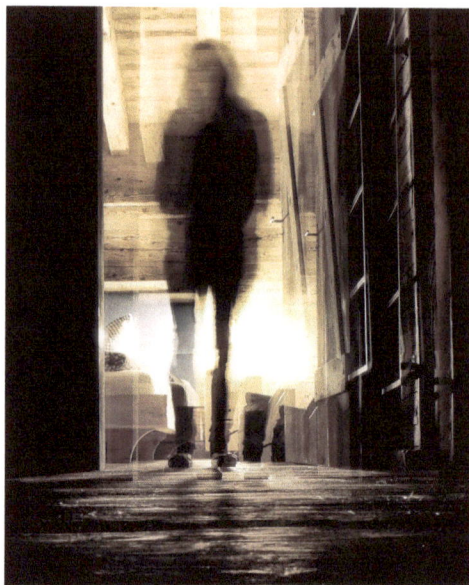

Person entering an alternate reality

In 1957, a Princeton physics graduate student named Hugh Everett showed that the consistency of quantum mechanics required the existence of an infinity of universes parallel to our universe. That is, there has to be a person identical to you reading this identical section right now in a universe identical to ours. Further, there have to be an infinite number of universes, and thus an infinite number of people identical to you in them.

So, is the multiverse a real possibility? There are some lines of thinking that point to an answer of "yes, it is a real possibility". In quantum mechanics, it is believed that when no one is looking at particles, they don't have set properties.

Instead, they are said to have all of the different possible properties that a particle could have. This is something called a wave function which is a mathematical description of the quantum state of an isolated quantum system. Once the particles are in a single universe, all those possibilities can't exist at once. So, when you look at a particle, it settles upon one single state and becomes that state.

Brian Greene, a physicist from Columbia University, explains another reason why the multiverse is a real possibility. He speculates that if space is infinite, it would mean that it must have infinite parallel realities. So, to simplify it, using a deck of 52 cards to represent the universe, and all of the matter in it, if you shuffle the deck enough times, eventually, the order of the cards you deal must repeat itself. Similarly, in an infinite universe, matter eventually would have to repeat itself, and arrange itself in similar patterns, thus creating alternate universes.

Most physicists, at least most physicists who apply quantum mechanics to cosmology, accept Everett's argument. So obvious is Everett's proof for the existence of these parallel universes, that Steven Hawking once said that he considered the existence of these parallel universes "trivially true." Everett's insight is the greatest expansion of reality since Copernicus showed us that our star was just one of many.

PART FOUR :

ALTERNATE UNIVERSES

7. THE ALTERNATE UNIVERSES

The European organization for nuclear research is known as CERN. There were 12 founding members: Belgium, Denmark, France, the Federal Republic of Germany, Greece, Italy, the Netherlands, Norway, Sweden, Switzerland, the United Kingdom, and Yugoslavia.

Physicists at CERN seek the answers to the question "What is the world made of?". They use some of the world's most powerful particle accelerators to do this. Some of their key achievements are the birth of the web, the Higgs Boson, and the Large Hadron Collider (LHC).

The LHC uses a 27-kilometer ring of superconducting magnets with a number of accelerating structures to boost the energy of the particles along the way. Two high-energy particle beams travel at close to the speed of light before they are made to collide. Thousands of magnets of different varieties and sizes are used to direct the beams around the accelerator.

It is located in a tunnel that is 100 meters underground, in the region between the Geneva International Airport and the nearby Jura mountains. The majority of its length is on the French side of the border.

CERN has compiled large quantities of data from these experiments which they stream to laboratories all over the world.

One of the quantum particles that CERN has been searching for is the graviton. Graviton particles are only hypothetical but it could give insight as to how gravity would react between different dimensions. The way CERN describes them is thought-provoking.

"If gravitons exist, it should be possible to create them at the LHC, but they would rapidly disappear into extra dimensions. Collisions in particle accelerators always create balanced events – just like fireworks – with particles flying out in all directions. A graviton might escape our detectors, leaving an empty zone that we notice as an imbalance in momentum and energy in the event. We would need to carefully study the properties of the missing object to work out whether it is a graviton escaping to another dimension or something else."

Some people think that these gravitons are creating holes to other dimensions and thus creating the Mandela effect.

Therefore, a question could be posed. Is CERN inducing these gravitons, creating holes to other dimensions and swapping idiosyncrasies in our world? Or, are we just having a collective memory lapse?

Alternate Universes presented as a schematic drawing

8. ADVENTURES IN AN ALTERNATE UNIVERSE

There are gateways or portals to a great many different worlds where things are very different from our own. One of the first people to popularize the idea of parallel universes was C. S. Lewis in his series "The Chronicles Of Narnia" where a wardrobe was a portal to an alternate reality.

Another author who wrote about parallel universes was Lewis Caroll. His heroine, Alice, entered a parallel universe by going down a rabbit hole.

Alternate realities have appeared in many science fiction stories and popular TV series such as Stargate, Star Trek, and Babylon V.

Steven Hawking once said that he considered the existence of these parallel universes "trivially true." This would be trivially true for most people, but for the people with vivid memories of an alternate reality, it is not so trivial.

The Man Who Showed Up At The Tokyo Airport

On a seemingly normal day in Tokyo in July 1954, a foreign man appeared at the Tokyo Airport. He told customs he was there on business. He presented everything needed to catch his flight including legitimate visa stamp and currency from multiple European countries. However, his cheques were from an unknown bank, and his driver's license had been issued in Taured. This of course was a problem because the country of Taured does not exist. The customs officials questioned him as to where Taured was, and he replied that it was on the border between France and Spain. When they looked on a world map they located the country of Andorra where the man said that Taured should be. This upset the man greatly, and he accused them of playing a practical joke on him. He insisted that his homeland had been there for thousands of years. He was clearly in shock that it was no longer on a map.

So, while the authorities sorted things out, the man was sent to a hotel to wait. Two immigration officers stood outside his hotel door.

The mystery deepened when they went to check on the man in the morning. He was not there. He had vanished. There was no way that the man could have gotten past the two immigration officers. His hotel room was 15 stories up above a busy street. He could not have jumped out without serious injuries or even death. The Metropolitan Police Department searched for him but never found him. If that wasn't strange enough, all of the man's documents, including his driver's license and passport that were issued in Taran, had also vanished from the security room of the airport. The man from Taured was never seen again.

Four Ladies Who Went Out For A Drive

Four ladies went out for a drive in the desert. They took a wrong turn and drove through a canyon. Then, they found themselves driving on concrete. They were surrounded by grain fields, and they saw a lake. They saw a building with a large neon sign. The sign seemed to have illegible random squiggles instead of words. The ladies stopped at a house for assistance. A large group of tall men came pouring out of the front door. They seemed terribly upset and were waving their arms at the women.

The women suddenly noticed that these tall men did not seem to be human. The women became terrified and drove off. As the women were driving away they saw four egg shaped automobiles mounted on tricycle style wheels following them.

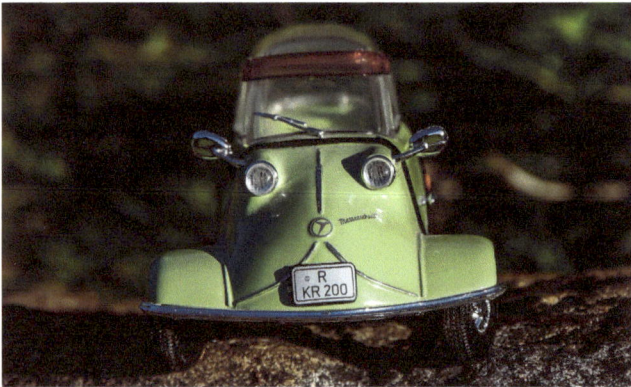

Egg shaped automobiles following them.

They sped up until the mysterious automobiles were out of sight. When they reached the canyon, they drove back through it. They had returned to the desert they were originally in. The ladies were unable to figure out the mysterious place where they had just gone nor how they had arrived there.

The Green Skinned Children Who Just Appeared

In the 12th century two children, a brother and a sister, appeared in the village of Woolpit, in Suffolk, England. It is said that they just popped up. They both had greenish coloured skin, spoke an unknown language, and wore bizarre clothes. They refused to eat any food except raw beans.

The children were taken in and cared for. Eventually they adapted, developed a taste for food, and lost the green hue in their skin. Shortly after, the boy died.

After learning to speak English, the girl explained that she and her brother were from a place where the sun did not shine bright, a place where it was perpetually twilight. She and her brother were out herding their dad's cattle when they heard a loud noise, and they were suddenly in a new place where they were discovered in Woolpit. The girl grew up and adapted herself to her new surroundings. It remained a mystery how she got there. The girl went on to get married and lived her life here in this dimension where she was not born.

The Woman Who Woke Up In The Morning With Her Life Slightly Changed

On July 16, 2008, Lerina Garcia, a Spanish woman, woke up in the morning and noticed minor differences in her surroundings. Strangely, her bed sheets were unfamiliar, and she did not remember putting on the pyjamas she was wearing. They were not the ones she had gone to bed in. She had no explanation for this, but she had to shake off the weird feeling it gave her, and she got ready for work.

Lerina went to work at the office she had been going to for 20 years. There she discovered that her office was no longer hers. Her name plate was missing and a man she had never seen before was at her desk. She thought at first that perhaps she had gotten the floors mixed up and was on the wrong floor. But after looking around she realized that she was on the correct floor. She found out that she did work in the same building, but in a different department, under a boss she had never met before. She began to get scared, so she told her boss that she felt ill, and she went home.

Upon her return home, she found that her ex-boyfriend was there — only he wasn't her ex, he was her current boyfriend. She clearly remembers breaking up with this guy and had been dating a new guy from her neighbourhood for about 4 months. When she tried to find the new boyfriend he was nowhere to be found. All traces of him were gone. She hired a detective to find him but for all intents and purposes, the new boyfriend had never existed.

Lerina believes she woke here in this dimension, but that she's from a parallel universe. The differences between the two universes are as small as her bedspread, and as significant as her love life.

The Cabin In The Markawasi Stone Forest

Markawasi Stone Forest is a plateau in the Andes mountains. The stones there have unique shapes and some look like faces. Many natives believe that there is a dimensional doorway in this stone forest.

A woman and her friends were camping at a site in the stone forest. They had gone out exploring the area at night when they happened upon a small stone cabin. They heard music coming from the cabin that was lit up by torch light. Inside, they could see people wearing 17th-century fashion dancing. The curious woman felt drawn to it, so she headed over and tried to enter. She was half-way in the door but before she could get inside, her friends pulled her out. Immediately, half of her body became paralyzed — the half that had entered the cabin.

A stone cabin

Her friends took her to seek medical care. Tests did not reveal any medical cause for her paralysis. Some believe that the woman partially entered a dimensional gateway, and when she was pulled out, she experienced a shift that threw off her nervous system, resulting in her partial paralysis. Had she not been pulled out by her friends, she may have been lost forever in another dimension.

The Universe In Which The Beatles Never Broke | Up

In 2009, James Richards was driving home from Turlock
with his dog. In Livermore, California, he stopped to allow his
dog empty her bladder. The dog sprinted after a rabbit.
Richards followed, tripped and was knocked unconscious.

When James woke up, he was next to an odd looking
machine. He was also with an old man named Jonas who
claimed that he had found Richards's unconscious body while
he was on a work trip for an inter dimensional travel agency.
The two started to talk about their respective pop culture.
Richards discovered that the Beatles were also present in this
dimension. But here they had never broken up and were all still
alive and were still making music. Richards purchased a cassette
tape labeled "Everyday Chemistry" which contained Beatles
songs which had never been heard in our reality. James
Richards uploaded the songs to his web page.
http://thebeatlesneverbrokeup.com

A Strange Man Shipwrecked In Germany

A man showed up in a village in Germany. The authorities picked him up and questioned him. He said that his name was Jophart Vorin. He spoke a broken variation of German. He said that he was from a part of the world known as Sakria. He was searching for his brother. The two had become separated during a shipwreck. He could not trace his route on shore. He spoke of other places such as Sakria, Aflar, Aslar, Auslar, and Euplar. He never found his brother.

He wound up living in Berlin

The scientists there were very curious about him and it seemed to them that he was from another planet.

Transported To A Parallel Universe While Driving

In November 1986, a man named Pedro Oliva Ramirez was making his usual journey home from Seville, Spain to his home in the town of Alcalá de Guadaíra.

He drove around a curve and found himself on an unfamiliar, highway that had six lanes. As he drove, he began to notice other unfamiliar things. He saw 20 story housing units, strange structures, odd flying air craft, and grass growing along the side of the road that was about two feet tall. A disembodied voice informed him that he had been transferred to another dimension. There was nothing for him to do but to continue driving.

He drove for about an hour before he decided to stop. He got out of the car to examine his surroundings. He noticed that all the cars that went by seemed to be hovering above the road and didn't have any tires. They passed by him at exactly 8 minutes apart. They were all vintage models and were beige or white in colour with darkened windows. Even the narrow, rectangle license plates were unusual. He got back in the car and started driving again.

Finally, after about an hour of driving, he spotted a name that was familiar to him. At a turn off he saw three signs pointing in three different directions. One was labeled Malaga, another was labeled Alcabala and the last was labeled Seville. Happy to see the sign that had a familiar destination written on it, Pedro took the Seville detour. At least if he could get back to where he started from in Seville then maybe he could find the correct way back to Alcalá de Guadaíra.

Eventually he stopped driving and when he got out of the car he was   surprised to find himself standing outside of his home in Alcalá de Guadaíra. He was now very confused. He tried to retrace his path but could never find the highway, the crossroads, nor the triple sign again.

PART FIVE :

CONSPIRACY

9. CONSPIRACY

Our world is full of communications. With so much communication and a human tendency to gossip, there is inevitably some lying and falsehood in these communications. Thus, some news and news casts are fake news. These false statements rise without any supporting evidence. Some false stories have been called conspiracy theories. These should however not be considered to be theories in the scientific sense.

As long as there has been politics, there has been conspiracies. In today's world one political party will state stories of what the opposing politician is up to. This is an effort to get voters to vote for their party ("because my party is more honest and transparent")

People may believe things which are not real even without conspiracy theorists attempting to make us believe a falsehood. Some of these things have their origins in our childhood when we were taught to believe in certain things. We outgrew a lot of our childhood beliefs. In some cases, we learned the hard lesson that our parents had actually lied to us. We had to discard Santa Claus, the Tooth Fairy, the Easter Bunny, and other fascinating beings. No wonder some children found it hard to believe that there is no Boogey Man. Is he under the bed or in my closet?

But what about the ones we did not outgrow? As we formed our own adult belief system we found some things easier to believe or disbelieve. Some people seem to find it easier to believe in false statements rather than investigate to find the truth.

The term theory is used when dealing with science. In science the stuff of the universe is studied such as astronomy, biology, quantum physics, physics. The observations in the studies are individual observations. The scientist then endeavors to explain all the observations on a topic. This is called a theory. An example is the theory of evolution. As more studies and experiments take place some results will not fit the theory. When that happens, the scientists change the theory to fit all the observations.

The dictionary defines theory as a supposition or a system of ideas intended to explain something, especially one based on general principles independent of the thing to be explained. This definition is the one used when discussing conspiracy theories and is quite different from the definition used in science.

10. CHILDREN'S TELEVISION

As we learn to talk, we also start to learn false information. Yes, Santa Claus arrives on Christmas Eve. The Easter Bunny hides coloured Easter eggs all over certain rooms in your home. (Nobody explained how he got in and out while avoiding your cat). Your tooth fell out and in the night the Tooth Fairy bought it by taking it from under your pillow and leaving a coin. But one day you discover somehow that your parents lied to you. Usually a confession was involved. By this time, there was television for the child to watch.

My first experience with television was when I was 10 years old. We lived in King City and had immigrated from Denmark. We had no TV of our own. We watched it at the neighbours house. There were obvious children's programs like the Howdy Doody Show, Tell A Story Time, and Walt Disney Cartoons. There was also a Superman series starring George Reeves. We knew that these programs were for entertainment and not to be taken as real life examples.

The big entertainment in movies at the time were the western movies and TV shows. The stars of these children's TV westerns included Johnny Mac Brown, Hopalong Cassidy, The Lone Ranger, Gene Autry, and Roy Rogers. There were serious outlaws in these episodes who robbed at gun point. There was usually a showdown between the outlaw and the hero. (The outlaw usually wore a black hat and the hero wore a white hat). The hero would usually win by shooting the gun out of the villain's hand. There was the draw and a shot that sent the gun flying out of the hand. In the process no hand was ever injured. These westerns were made so that children were shielded from violence which was very good and provided role models.

Roy Rogers and Dale Evans with Roy's horse, Trigger

These westerns did not affect our belief systems very much. However, other shows did affect our belief systems in a sometimes not so subtle way. One of these examples were the jungle movies. These involved a group of European people going into the jungle for various reasons such as hunting, finding treasure, finding lost cities, etc.

The jungle movies would show a group of people going to Africa. The usual preamble had them going to the Darkest Africa. No real location was given. The group would arrive at an African village where they hired guides and porters. Then, they headed into the jungle in single file with the guide cutting a trail using a machete.

As the file of people walked along the trail they saw several animals and birds. There were always monkeys shown. They always encountered animal attacks from large cats. In Africa, it was usually a lion which they had to kill to save themselves. The lion was portrayed as the most dangerous animal in the jungle. Therefore, it became known as the King of the Jungle.

King of the Jungle

For years my mind told me that the lion was the king of the jungle. Then, I discovered that lions do not live in jungles. There was more to Africa than the jungle. Lions live in grass lands; the savanna. Walt Disney recognized this and his latest movies had the lion not as the "king of the jungle" but it was recognized as "The Lion King".

As children develop, they adopt their own belief systems. These start with the belief systems of their family. As they develop and go out into the world they add to their own belief systems. Because of falsehoods in the world, false beliefs may also be incorporated into their belief system. This may sometimes lead to a belief in one or more conspiracy theory.

11. AREA 51 AND UFOS

In July 1947 I was 5 years old and my family was visiting my father's parents, Laurits and Johanne Jorgensen. One day, all my father's brothers were there, Gordon, Richard, Borve, Peter, Svend Aave, and Evald. Svend Aave was drawing pictures of cars. He was very good at it. The cars were very life like.

Somebody turned on the radio to listen to the news. There was an announcement that a flying saucer had crashed in the USA in New Mexico. My mind envisioned a dinner plate falling to the ground. After listening to the news for awhile it became clear to me that a flying saucer was no dinner plate but a good sized flying craft. The term UFO was never mentioned. The object was referred to as a flying saucer.

The flying saucer held occupants; aliens of a type no one had seen before. The aliens were reported being dead and personnel from a secret U.S. Air base had transported them to their air base.

On June 14, 1947, William Brazel, a foreman working on the Foster homestead noticed clusters of debris approximately 50 kilometers north of Roswell, New Mexico. He paid little attention to it. He returned to the site on July 4th with his wife, son, and daughter. They gathered a lot of the debris and carried it home.

The next day, Brazel heard reports of a flying disc and wondered if that was what he had carried home. On July 6 Brazel went to Sheriff Wilcox and whispered to him that he may have found a flying disc.

Wilcox called RAAF Major Jesse Marcel. A man in plain clothes then accompanied Brazel back to the ranch where he gathered up all the debris. On Monday, July 7 they spent a couple of hours looking for more parts and found a couple more patches. The plain clothed man carried away all the debris.

Early on Tuesday, July 8, the RAAF issued a press release. It was immediately picked up by numerous news outlets.

"The many rumors regarding the flying disc became a reality yesterday when the intelligence office of the 509th Bomb group of the Eighth Air Force, Roswell Army Air Field, was fortunate enough to gain possession of a disc through the cooperation of one of the local ranchers and the sheriff's office of Chaves County. The flying object landed on a ranch near Roswell sometime last week. Not having phone facilities, the rancher stored the disc until such time as he was able to contact the sheriff's office, who in turn notified Major Jesse A. Marcel of the 509th Bomb Group Intelligence Office. Action was immediately taken and the disc was picked up at the rancher's home. It was inspected at the Roswell Army Air Field and subsequently loaned by Major Marcel to higher headquarters."

Colonel William H. Blanchard, commanding officer of the 509th contacted General Roger M. Ramey of the Eighth Air Force in Fort Worth, Texas. Ramey ordered the object to be flown to Fort Worth Air Field. At the Air base, Warrant Officer Irving Newton identified the object as a weather balloon and its kite, confirming Ramey's original opinion. "Kite" is a nickname for a radar reflector used to track the balloon from the ground.

Later that day, the press reported that the Commanding General of the Eighth Air Force, Roger Ramey, had stated that a weather balloon had been recovered by the RAAF personnel. A press conference was then held featuring crash debris said to be from the weather balloon. This achieved the desired effect of squashing the story. The military had also concealed the true purpose of the crashed device – nuclear test monitoring.

There are people who study UFO phenomenon. They are called ufologists. Between 1978 and the early 1990s interviews with hundreds of people were conducted by ufologists, Stanton T. Friedman, Willian Moore, Karl T. Pflock, Kevin Randle, and Donald R. Schmitt. They interviewed people who claimed to have ties to Roswell in 1947. Using the Freedom of Information Act requests, hundreds of documents were obtained along with other documents such as Majestic 12 which had been leaked by insiders (supposedly). The ufologists concluded that at least one alien space craft had crashed near Roswell, alien bodies had been recovered, and a government cover-up of the event had taken place.

Over the years, there were books, articles, and TV specials about the Roswell flying saucer. In a 1997, a CNN/Time poll revealed that the majority of the people interviewed believed that aliens had visited the Earth and that all relevant information was being kept secret by the U.S. Government.

The Roswell incident inspired the movie "Independence Day" and the TV series "The X Files".

Today, there are disagreements about the Roswell incident. Was there only a weather balloon crash or did an alien space craft also crash there?

UFO over a farm near McMinnville, Oregon

SIGHTINGS IN CHINA

A few years after the Area 51 incident, there was a news broadcast about a flying saucer which had crashed in a mountain in China. Nothing more was ever reported about this particular crash.

In 1994, Meng Zhaoguo and a relative saw what they thought was a weather balloon descend into the Red Flag forest in Hellongjiang province. The two followed the crashing balloon into the forest and looked at the shining object. After this encounter, Meng claimed that he was continually harassed by entities from the flying saucer. He said that he was taken to their space ship and forced to copulate. On July 17, he was abducted from his house, taken into the space craft and shown a planet which the entities said was their home world. Meng thought he was being shown Mars.

In 1997, his story was analyzed by the UFO Enthusiasts Club at Wuhan University. They came to the conclusion that while the original contact may have happened, the successive described events were in all likelihood false. Notwithstanding, other UFO groups in China inferred that his reports were true.

Hangzhou Airport

At eight o'clock in the evening on July 10, 2010, an unidentified flying object was spotted hovering above Hangzhou airport. The airport officials received orders to shut down the airport until the sky was cleared. All outgoing planes were kept on the ground and incoming planes were redirected to the Ningbo and Wuxi airports that were nearby. About two hours later normal operations were resumed at Hangzhou airport.

SIGHTINGS IN CANADA

Kingston, Ontario

In 1962, I was a student at Queen's University in Kingston. I was living at the student residence called Morris Hall which is located near Lake Ontario. From Morris Hall you can see the lake and a path running alongside of it. The path was just across the street from Morris Hall. One day, as I was walking back to my residence after attending some classes, I saw a strange light moving over the lake. Later that evening, I was watching a news broadcast in which it was stated that UFOs had been seen over Lake Ontario. It was also stated that there had been an unusual amount of UFO sightings over Lake Ontario near Kingston in the recent past.

Air Canada Flight 305

In 1967, Air Canada Flight 305 was en route to Toronto from the Halifax International Airport. At 7:15 p.m., when over Sherbrooke and St. Jean Quebec, the pilot, Captain Pierre Charbonneau, spotted something and pointed it out to co-pilot, Bob Ralphington. On the left side of the plane an object was tracking along on a parallel course a few miles away. The object was rectangular and brilliantly lit. There was a string of smaller lights trailing the object. At 7:19 the pilots noticed a considerable noiseless explosion near the rectangular object. Two minutes later, a second explosion occurred which faded to a blue cloud around the object.

Mahone Bay, Nova Scotia

While sitting on their front porch in Mahone Bay, Darrel Dorey, his sister Annette, and their mother watched a large yellow object maneuvering in the southwestern horizon. The next day Darrel wrote a letter to the RCAF Greenwood Base Commander asking what had been flying over the water that evening because he had never seen anything like it.

Sambro, Nova Scotia

Captain Leo Howard Mersey was standing in his vessel's wheelhouse. He was watching four blips on his Decca radar. The blips were stationary. Then, he looked up and saw four bright objects hovering about 28 km from the vessel. They were roughly in a rectangular formation.

The entire crew of twenty fishermen stood on the deck and

watched the objects in the northeastern sky. Mersey radioed the rescue co-ordination center and the harbour master in Halifax and asked for an explanation. When he arrived in port, he filed a report with the Lunenburg RCMP.

Shag Harbour, Nova Scotia

On October 4, 1967, at around 10 p.m., the Chronicle Herald and local radio stations reported that many people had called to report strange glowing objects flying around Halifax. At 11:20 p.m., it was reported that something had crashed into Shag Harbour. Eleven people had seen a low flying, lit, object crash into the water. They reported hearing a whistling sound, like a bomb, then a loud "whoosh", and finally a loud bang. It was never officially identified and was referred to as an unidentified flying object (a UFO). A local resident, Laurie Wickens, and four of his friends reported the incident. They had seen a large object floating 820 to 900 feet offshore in Shag Harbour. Wickens contacted the RCMP in Barrington Passage.

About 15 minutes after the call by Wickens, two RCMP officers arrived concerned for survivors. The RCMP detachment called the Rescue Co-ordination Center in Halifax to advise them of the situation and ask if any aircraft were missing. The object then started to sink and disappear from view.

A rescue operation was amassed within a half an hour. Local fishing boats went out to look for survivors. Nothing was found by either the fishermen or the Canadian Coast Guard search and rescue team. There were no survivors, no bodies, nor any debris.

By the next morning, RCC had determined that no aircraft were missing. No commercial, private, or military aircraft were missing. Along the eastern seaboard in both Atlantic Canada and New England, all aircraft were accounted for.

RCC Halifax also sent a telex to the "Air Desk" at the Royal Canadian Air Force in Ottawa informing them of the crash and that all conventional explanations such as aircraft, flares, etc. had been dismissed. Therefore, this was labeled as a UFO report.

Fleet Diving Unit Atlantic did an underwater search. Divers combed the sea floor of the Gulf of Maine off Shag Harbour looking for an object. The final report stated that not even a remnant of the object was ever found.

The Chronicle-Herald ran a story on October 9 titled "UFO Search Called Off". This was also carried by the Canadian Press in other Canadian newspapers.

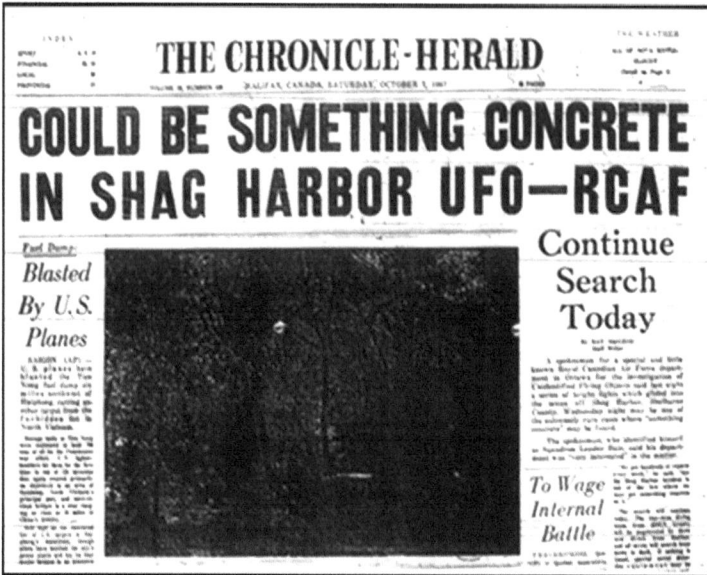

Halifax Chronicle-Herald reports the 1967 Shag Harbor UFO sighting

The Baltic Sea Anomaly

In 2012, on a dive to search for an old ship wreck, Swedish explorer, David Lindberg, and his Ocean X team of explorers found a 200 ft wide object 300 ft below the surface in the Baltic Sea off the coast of Sweden, between Sweden and Finland. It strongly resembled Han Solo's Millennium Falcon and had a staircase leading to a dark hole.

The team contacted geologists and marine biologists. At first, it was considered a natural object. Speculations considered that the object might be led some to say that it might be an anti-submarine device left over from World War Two, or a battleship gun turret, or indeed a flying saucer.

Despite looking like a huge rock, the scientists determined that it was made of metal. When the scientists tried to get close to the object, electric and satellite equipment cut out. Stefan Hogerborn, a professional diver with Ocean X, said, "Anything electric out there, and the satellite phone as well, stopped working when we were above the object. Then, when we got away about 200 meters, it turned on again."

The object was aged as being 130,000 years old.
For five years Mr. Lindberg's Ocean X team were investigating the object.

Surveys Of UFO Sightings

The Canadian UFO surveys come from Winnipeg. A survey stated that the year with the most UFO reports was 2012. In that year there were 1,981 reports. In 2015 there were 1,267 reports. Quebec was responsible for 35% of all these reports. In the past decade it was responsible for 5 to 15%. Thus reports from Quebec had gone way up in 2015.

There was an increase in UFO sightings in 2015 in the provinces of Alberta, Saskatchewan, Manitoba, Nova Scotia and Newfoundland. There were 97 reports in Montreal, 78 in Toronto, 69 in Vancouver, and 36 in Edmonton. The study said that a typical sighting lasted about 16 minutes. Most of the sightings were of simple lights in the sky. The continued reporting of UFOs by the public and the continued increase in numbers of UFO reports suggests a further examination of the phenomenon by social, medical, and/or physical scientists are needed.

12. NASA FAKED THE MOON LANDING

On July 20, 1969, on a TV at a friend's house, my parents and my wife Brenda watched the lunar module from the Apollo 11 space craft land on the moon. Neil Armstrong (commander), Buzz Aldrin (lunar module pilot), and Michael Collins (command module pilot) were the crew. We watched Neil Armstrong step on the moon and hear him say, "One small step for man. One giant step for mankind."

Buzz Aldrin on the moon

The crew got out an American flag and displayed it for the camera.

American flag displayed on the moon

In 1977, a movie about a space mission was released called "Capricorn 1". The movie showed a fake mission with the sequence of events being filmed in a studio.

The plot of the movie was that the space module, which was shot into space by NASA, exploded at some point. The people in charge of the project, realizing that the people watching the event would think that the astronauts were killed in the explosion, decided to kill the astronauts involved so that the illusion of them having gone into space would not be detected. In the movie, O. J. Simpson's character escaped from the assassinations and wound up in one of the American deserts where he was pursued by teams of assassins.

After this movie, conspiracy theorists have claimed that the moon landing was a hoax and was staged in a film studio.

In 2001, the Fox network aired a documentary called "Moon Landing Hoax". The program stated and illustrated the following:

1. You cannot see the stars in pictures taken by Apollo astronauts. Without air, the stars should be more visible.

2. The shadows in the Moon pictures are cast in different directions. This means there must be at least two light sources. The pictures appear to be backlit.

3. There is no blast crater under the Lunar Module. The Lunar Module should have created a blast crater as it landed.

4. The Lunar Module should have kicked up dust as it landed. That dust should be visible on the feet of the Module.

5. As the flag is deployed, it waves due to a rogue breeze. There is no breeze on the moon, so this must have been filmed on Earth.

6. The Van Allen Belts (the radiation belts above Earth) would kill any astronaut passing through it.

7. The "Moon rocks" were meteorites from Antarctica or were manufactured in a lab.

Eventually, a film was made debunking the theory that the moon landing was a fake.

#1 Where are all the stars? - **Y**ou cannot see the stars in pictures taken by Apollo astronauts. Without air, the stars should be more visible.

The answer to this question has to do with how cameras work. Since the light of the sun was very bright and the astronauts were brightly lit from that sunlight, the cameras could not be set to have long exposure times or the pictures would be all washed out. Instead, the cameras were set up to take pictures at short exposure times which did not allow enough time for the stars to be exposed on camera.

Another example of the low exposer rate of the camera is seen in an image that was taken by an astronaut leaving the International Space Station on a Russian Soviet spacecraft. It shows the Space Shuttle Endeavour docked to the International Space Station (ISS) during its final mission on May 23, 2011. It was taken with a low exposure rate making the stars invisible.

The space shuttle Endeavour docked to the ISS

#2 Stage Lighting - The shadows in the Moon pictures are cast in different directions. This means there must be at least two light sources. The pictures appear to be back lit.

The moon hoax people state that anyone who has walked outside at night under street lamps will have noticed many shadows cast by different street lamps. One light source equals one shadow. If there were multiple light sources, you would see many shadows. These people were referring to shadows cast in different directions. On the pictures there is only one shadow per astronaut. The reason the shadows appear in different directions is an effect caused by the uneven lunar surface and the two-dimensional nature of the photographs. As for the pictures appearing to be backlit, the sun is not actually the only source of light. Well, technically it is but the overwhelming reflection of that sunlight as it hits the surface of the moon. Think of it like trying to stare at the bright white snow or at the white sands of a beach on a sunny day. There is so much light that it hurts your eyes. It is that kind of intense reflection of the sun on the surface of the moon that causes the pictures to appear as though they are back lit.

#3 No blast crater - There is no blast crater under the Lunar Module. The Lunar Module should have created a blast crater as it landed.

When landing, the lunar module exerted only 1.5 pounds per square inch. That is not very much; certainly not enough to produce a blast crater.

\# 4   The Lunar Module should have kicked up dust as it landed. That dust should be visible on the feet of the Module.

The module did kick up a little dust on the moon but not very much. This was because there is no air on the moon. When a rocket blasts off from the Earth, air pressure helps whip everything around. On the moon, only dust that touched the exhaust was displaced.

Lunar lander

#5 As the flag is deployed, it waves due to a slight breeze. There is no breeze on the moon, so this must have been filmed on Earth.

The flag was held in position by both horizontal and vertical bars. The waving was not caused by a breeze. It was caused by the astronauts twisting the pole to make the flag stand upright. The bottom bar did not extend properly which made the flag uneven and simulates the look of a ripple in a breeze. NASA not only liked the look of the way this flag was displayed, but they purposely designed subsequent flags to look the same on following missions.

The flag was twisted to make it stand upright

#6 The Van Allen Belts (the radiation belts above Earth) would kill any astronaut passing through it.

    The Van Allen belts are areas above the Earth where particles from the solar wind are trapped in Earth's magnetic field. If a man were to stay there unprotected for a long enough period, he would surely die. The space ship passed through them quickly and was through in about an hour. There was not enough time to get a lethal dose of radiation. As well, the metal hull of the space ship blocked most of the radiation.

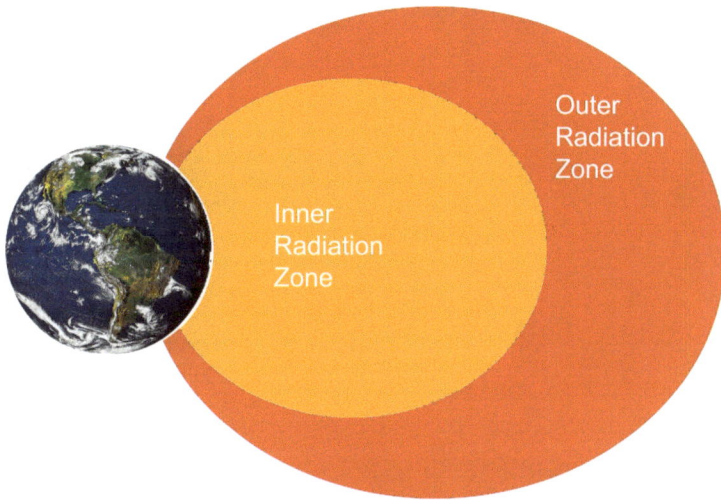

Earth showing Van Allen belts

#7 The "Moon rocks" were meteorites from Antarctica or were manufactured in a lab.

Moon rocks

All meteorites get scorched and oxidized as they fall through Earth's atmosphere and land on the ground. These moon rocks did not have any scorch marks or oxidization. The moon rocks were also tested against meteorite rocks from Antarctica. The tests proved that they were from the moon.

The moon landing was real!

13. REPTILIAN HUMANOIDS

UFO sightings have been recorded in history from as far back as 1440 BC. All through our planet's history since 1440 BC there have been UFO observations. Recently most of these observations have lasted for about 15 minutes or longer.

Some people, such as Eric Von Daniken and Zecharia Sitchin, have maintained that aliens have landed on earth in the past and have played a part in our history by being associated with our ancient gods. Some people (who may be considered by others as conspiracy theorists) claim that aliens are still with us on our planet.

The theorists call these aliens reptilian elite. They are described as blood-drinking, flesh-eating, shape-shifting extraterrestrial reptilian humanoids. The theorists say that they are among our leaders, our corporate executives, our beloved Oscar-winning actors and Grammy-winning singers. They also say that these reptilian elites are responsible for the Holocaust, the Oklahoma City bombings and the 9/11 attacks.

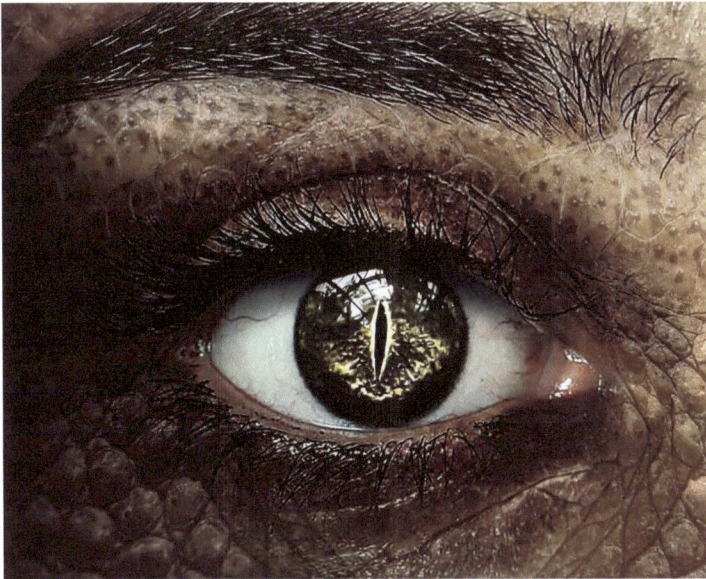

The former BBC sports reporter, David Icke, published his first book in 1998 called "The Biggest Secret". In his book were interviews with two British people who allege that members of the royal family are just reptile aliens with crowns.

Icke claimed that the reptiles (known as the Annunaki) have had humans under their control since ancient times. Purported to be among their number are Queen Elizabeth, George W. Bush, Henry Kissinger, Bill and Hillary Clinton, and Bob Hope. Icke additionally asserts that the reptilians are behind the mystery social orders like the Illuminati and the Freemasons. This reptilian conspiracy theory inspired the TV series "V".

Conspiracy theorists claim that the government is covering up their knowledge of aliens. Stephen Hawking has stated, "If the government is covering up knowledge of aliens, they are doing a better job of it than they do at anything else."

14. SECRET SOCIETIES

Secret clubs peaked in the 18th and 19th centuries. These secret societies served as safe havens to discuss everything from science, academia, and religion. It was needed to avoid the scrutiny of the church and state at that time. Because the societies were secret, people became very distrustful of them.

In 1695, Count Anthony Von Sporck started a society called the International Order of St. Hubertus. It was meant to be a society for elite hunters. The Order of St. Hubertus denied membership to the Nazis. The military leader, Herman Goering, was thus denied membership. Because of that, Hitler dissolved the Order. The Order emerged again after World War II. An American chapter was founded in the late 1960s.

Today, there are many clandestine organizations. The New York Times has written articles about them. Conspiracy theories about these organizations are believed by many people. Freemasons have been a focal point of these conspiracy theories.

Such theories include the following:

- Freemasonry brought about the Civil War.

- Freemasonry has acquitted President Johnson.

- Freemasonry has committed or concealed crimes without number.

Many believe that European secret societies are the ruling power in Europe.

Many religious leaders felt very conflicted about secret orders. In 1887, Reverend T. De Witt Talmage wrote his sermon on "the moral effect of Free Masonry, Odd Fellowship, Knights of Labor, Greek Alphabet and other Societies." Cardinal James Gibbons stated that the secret orders had no excuse for existence.

In the United States in the late 19th century, there was such an uproar against secret societies that one concerned group created an annual "Anti Secret Society Convention". In 1869, at the annual convention in Chicago, the organization's secretary stated that the secular press either approved or ignored secret societies. The secretary also stated that few religious papers had spunk enough to come out for Christ in opposition to Freemasonry.

The novelist, Dan Brown (author of The Da Vinci Code), and his contemporaries have mentioned some of the bigger organizations such as the Order Of Skull and Bones, Freemasons, Rosicrucian, and the Illuminati.

The Order of Elks was formed in Cincinnati, Ohio and was called The Benevolent and Protective Order of Elks of the World. In 1899, two black men were denied membership to the organization. The two men decided to form their own organization. They called it The Improved Order of Elks of the World. This Order was once considered to be the center of the Black community. During the age of segregation, the lodge was one of the few places where black men and women could fraternize. The secret society continues to sponsor educational scholarship programs, youth summer computer literacy camps, parades, and community service activities throughout the world.

The Grand Orange Lodge also known as the Orange Order got its name from Prince William III; the Prince of Orange. It was founded after the Battle of the Diamond in a small village called Loughgall in modern-day Northern Ireland. Its purpose was "to protect Protestants". At that time, secret societies were banned from Ireland. The Lord Lieutenant of Ireland, George William Frederick Villiers supported the Grand Orange Lodge. The Dublin Waterford News openly criticized the Lord Lieutenant. The paper wrote, "Lord Clarendon has been holding communication with an illegal society in Dublin for upwards of ten days. The Grand Orange Lodge, with its secret signs and passwords, has been plotting with his Excellency during all that time. It may seem strange, but it is fact."

The Grand Orange Lodge is still around in Ireland. It also has clubs all around the world. Many people believe that they are anti-Catholic, but their official website states "Orangeism is a positive rather than a negative force. It wishes to promote the Reformed Faith based on the Infallible Word of God - the Bible. Orangeism does not foster resentment or intolerance. Condemnation of religious ideology is directed against church doctrine and not against individual adherents or members."

Only members of the Order know when the Independent Order of Odd Fellows first started. Before he was named Prince Regent of the United Kingdom, George IV had been a member of the Freemasons. In 1812, he wanted one of his relatives to be a member without having to endure the lengthy initiation process. His request was emphatically denied. According to a history of the Independent Order of Odd Fellows published by the Philadelphia Evening Telegraph in 1867, George IV left the Freemasons and declared he would develop a rival club. However, the official website of the Order of Odd Fellows traces its beginnings all the way back to 1066.

The Independent Order of Odd Fellows still exists today. They call themselves the Odd Fellows, and they are grounded in friendship, love, and truth. Their members included Winston Churchill and Stanley Baldwin.

The Knights of Pythias was founded by Justus H. Rathbone,

an employee of the government in Washington D.C, in 1864. He felt there was a need for an organization that practiced brotherly love since the Civil War was going on at the time. It was the first order to be chartered by an act of Congress. The Knights of Pythias is still active and is a partner of the Boy Scouts of America.

The Ancient Order of the Foresters was rooted in England in 1834. It was established before State Health Insurance was put into place in England. The club offered sick benefits to its working class members.

In 1874, the Canadian and American branches left the Order and set up the Independent Order of the Foresters. Candidates who wanted to join the club had to pass an examination by a competent physician. The society still provides insurance policies for its members and engages in a variety of community service activities.

In 1868, the Ancient Order of United Workmen was formed by John Jordan Upchurch and 13 others in Meadville, Pennsylvania. Their goal was bettering the conditions for the working class. Just like the Foresters, it set up protections for its members. Should a member die, all brothers of the order contributed a dollar to the member's family. It was eventually set to a maximum of $2,000.00. The Ancient Order of Workmen is no longer around, but it had influenced other Orders to add an insurance package to its members.

The Patriotic Order Sons of America was established back in the early days of the American Republic. It was said to be one of the "most progressive, most popular, most influential, as well as strongest patriotic organizations" in the United States in the early 20th century. However, in 1891 it refused to delete the word white from its constitution and did not allow black men to apply. This has changed. Today, the order opens its membership up to "all native-born or naturalized American male citizens, 16 years and older, who believe in their country and its institutions, who desire to perpetuate free government, and who wish to encourage a brotherly feeling among Americans, to the end that we may exalt our country, to join with us in our work of fellowship and love."

The secret society called the Molly Maguires was brought to the United States by Irish Immigrants. The Maguires got its name because members dressed in women's clothes to carry out illegal activities such as arson and death threats. In the 1870s, 24 foremen and supervisors were assassinated in the coal mines of Pennsylvania. It was suspected that this was carried out by the Molly Maguires. The Order of the Sons of St. George was another secret society established to oppose the Molly Maguires. The mining companies hired the Pinkerton Detective Agency to investigate the Molly Maguires. They planted a mole in the group. In a series of criminal trials, 20 Maguire members were sentenced to death by hanging. The Molly Maguires and the Sons of St. George have both disbanded.

"THE MARCH TO DEATH," depicts Molly Maguire members on the way to the gallows in Pottsville,Pennsylvania.(Corbis)

15. THE J. F. K. ASSASSINATION

In 1963, I was studying at Queen's University in Kingston, Ontario. I was living in the residence called Morris Hall. On the evening of November 22, I was in the common room watching news on TV with some of the other residents. The news of the John F. Kennedy assassination came on with a movie clip of the event. The movie clip showed the president and his wife in a cavalcade.

The sequence of events showed the president being hit by gunfire. Apparently one shot missed the presidential limousine entirely. A second shot hit both Kennedy and Texas Mayor, John Connally. The third bullet killed Kennedy when it struck him in the head. This shot sprayed blood and brain tissue on Connally. Connally was taken to a hospital where he went through four hours of surgery. The extent of his injuries were a punctured lung, three broken ribs, a shattered wrist, as well as a bullet lodged in his leg. Connally said, "There were either two or three people involved, or more, in this – or someone was shooting with an automatic rifle".

The news report said that the shots had come from a grassy knoll located just off the route of the presidential cavalcade.

About 45 minutes after the assassination, a man named Lee Harvey Oswald shot and killed Dallas police Officer, D. Tippit on a local street. Tippit had stopped Oswald to question him. After he shot Tippit, Oswald then went into a movie theater. There, he was arrested for Tippit's murder. Later, he was also charged with the assassination of the president. He denied both accusations and said that he was a patsy. Oswald was formally arraigned on October 23 for the murders of John Fitzgerald Kennedy and Officer J. D. Tippit.

The next day, Oswald was brought to the basement of the Dallas police Headquarters in preparation for transport to a more secure place. With cameras rolling, a crowd of news people and police gathered to witness his departure. As Oswald came into the room, a man named Jack Ruby sprung out of the crowd and killed him with a .38 revolver. He had hidden on him. He claimed he had done this because he was enraged by Kennedy's murder. Ruby operated strip clubs and dance halls in Dallas. He also had minor connections to organized crime.

The basement of the Dallas police Headquarters

Jack Ruby shot Lee Harvey Oswald

Since he was dead, Oswald could not be brought to trial. It was impossible to obtain a motive for the assassination. All through history important people had ordered assassinations which were covered up by having the assassin killed. Could this have happened here? People now started to wonder about that and conspiracy theories arose.

Shortly after the assassination, Lyndon B. Johnson was sworn in as President of the United States.

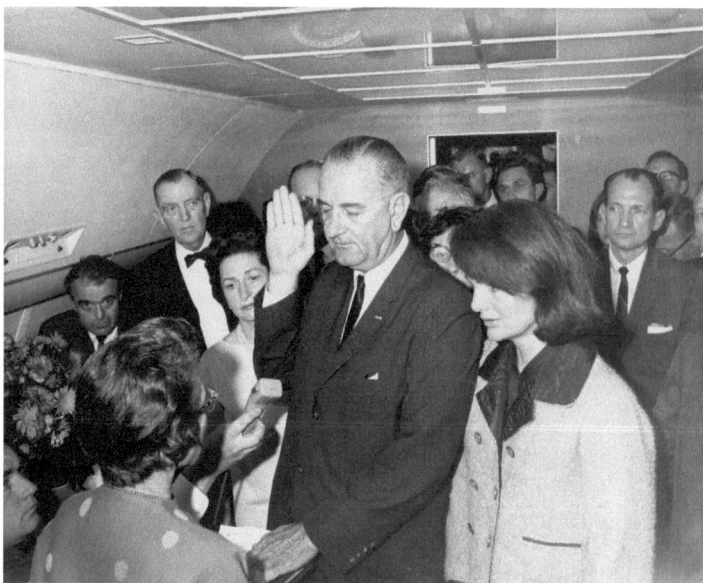

Lyndon B. Johnson being sworn in as President. Jacqueline Kennedy is on his left.

After Oswald was shot dead, Hoover, the director of the FBI, wrote a memo which stated that the Dallas police would not have enough evidence against Oswald without the FBI's information. He stated that he and the Deputy Attorney General were concerned about having something issued to convince the public that Oswald was the real assassin.

CIA intercepts had discovered that Oswald had visited the Cuban and Russian embassies in Mexico several weeks before the assassination. Special agents of the FBI who had interviewed Oswald in Dallas observed the pictures from Mexico and listened to his voice on tape. They stated that this person was not Oswald but an impersonator.

Hoover, in talking to President Johnson stated, "No, there's one angle that's very confusing for this reason. We have up here the tape and the photograph of the man at the Soviet Embassy, using Oswald's name. That picture and the tape do not correspond to this man's voice, nor to his appearance. In other words, it appears that there was a second person who was at the Soviet Embassy."

The Attorney General, Nick Katzenbach, wrote, "Speculation about Oswald's motivation ought to be cut off, and we should have some basis for rebutting thought that this was a Communist conspiracy or (as the Iron Curtain press is saying) a right–wing conspiracy to blame it on the Communists."

He also wrote, "The public must be satisfied that Oswald was the assassin; that he did not have confederates who are still at large; and that the evidence was such that he would have been convicted at trial"

Four days after Katzenbach's memo President Johnson formed the Warren Commission with Earl Warren as chairman.

THE WARREN COMMISSION

On November 29, 1963, President Lyndon B. Johnson established the Warren Commission through executive order 11130 to investigate the assassination of U. S. President John F. Kennedy that had taken place days earlier on November 22, 1963.

The U. S. congress passed the Senate Joint Resolution 137 authorizing the presidential appointed commission, led by Chief Justice Earl Warren, to report on John F. Kennedy's assassination; mandating the assembly and gathering of evidence and testimony of witnesses.

The Warren Commission

After nearly a year long investigation, the commission concluded that alleged gunman, Lee Harvey Oswald, had acted alone in assassinating America's 35th president and that there was no conspiracy either domestic or international involved.

"The 46-year-old Kennedy was shot while traveling in a motorcade in an open-top limousine as it passed the Texas School Book Depository Building in downtown Dallas at approximately 12:30 p.m. First Lady, Jacqueline Kennedy, Texas Governor, John Connally, and his wife, Nellie, were riding with the president. The governor also was shot and seriously wounded. Kennedy was pronounced dead 30 minutes later at Dallas Parkland Hospital."

The Warren Commission reviewed reports by the FBI, Secret Service, Department of State, and the Attorney General of Texas. They also studied Oswald's personal history, political affiliations, and military record. They listened to testimonies of hundreds of witnesses. They also visited the site of the assassination many times.

On September 28, 1964, the Warren Commission presented its 888 page report to President Johnson. This was released to the public three days later. The Commission concluded that the bullets that killed Kennedy and injured Connally were fired by Oswald from a rifle pointed out of a sixth-floor window in the Texas School Book Depository. Details of Oswald's life were described. This included a visit Oswald had made to the Soviet Union. No endeavour was made to establish his motives. The Commission concluded that the Secret Service did not make sufficient preparations for Kennedy's visit to Dallas and had failed to adequately protect him. They also concluded that Jack Ruby had acted alone in killing Oswald.

The newspapers then carried stories about the assassination. It included a picture of Lee Harvey Oswald holding his assault rifle. They used this picture in order to convince the public that Oswald was a lone killer.

Lee Harvey Oswald holding his assault rifle

The Warren Commission's report that Oswald was a lone gunman failed to satisfy some who had witnessed the attack. Others who did research found conflicting details in the Commission's report. Therefore, a number of conspiracy theories arose. Some involved the Cuban and Soviet governments. Some theories involved organized crime, the FBI and the CIA. Some even involved President Johnson.

Research

Numerous researchers, Mark Lane, Henry Hurt, Michael L Kurtz, Gerald D. McKnight, Anthony Summers, and Harold Weisberg have pointed out inconsistencies, oversights, exclusion of evidence, errors, changing stories, or changes made to witness testimony in the Warren report.

Michael Benson wrote that the Warren Commission was only given information furnished by the FBI, and that its goal was to make the lone gunman theory the only plausible theory.

United States Senator and U.S. Senate Select Committee on Intelligence member, Richard Schweiker, told author, Anthony Summers, in 1978 that he "believed that the Warren Commission was set up at the time to feed pabulum to the American public for reasons not yet known, and that one of the biggest cover-ups in the history of our country occurred at that time"

In 1998, Federal Judge and Assassination Records Review Board (ARRB) chairman, John R. Tunheim, stated that no "smoking guns" indicating a conspiracy or cover-up were discovered during their efforts to declassify documents related to the assassination in the early 1990s.

James H. Fetzer took issue with this statement. He identified 16 "smoking guns" that he claims proves the original narrative is impossible, and therefore, a conspiracy and cover up had occurred. He also claims that the evidence presented by the ARRB substantiates these concerns. The evidence in question included things like problems with the bullet trajectories, questionable findings about the murder weapon and the ammunition used, differences between the Warren Commission account and the autopsy findings, inconsistencies between the autopsy conclusion and what was reported by witnesses at the scene of the assassination. As well, there were eye witness accounts that conflicted with x-rays taken of the President's body, signs that the diagrams and photos of the President's brain in the National Archive do not belong to the President, testimony by those who took the autopsy photos that the photos were altered, created, or destroyed, indications that the Zapruder film had been tampered with, indications that the Warren Commission's version of events conflicts with news reports from the scene of the murder. Also, a suspicious change to the motorcade route that would have aided in the assassination, supposed lax Secret Service and local law enforcement security, and statements made by people who claimed they had knowledge of or participated in a conspiracy to kill the President.

Richard Buyer, author of "Why The JFK Assassination Still Matters", wrote that many witnesses were either ignored or intimidated by the Warren Commission. In "The Last Dissenting Witness" a 1992 biography of Jean Hill, Bill Sloan wrote that she was abused by Secret Service agents, harassed by the FBI, and received death threats. Jean Hill was the witness closest to the President's limousine when the shooting happened.

Several assassination eye witnesses report that their testimonies were cut short or stifled any comments casting doubts on the conclusion that Oswald had acted alone. Some of those witnesses include Richard Carr, Acquilla Clemens, Sandy Speaker, and A. J. Millican Marrs. An employee of the Texas School Book Depository, Joe Molina, was browbeaten by authorities. He lost his job soon after the assassination. Witness, Ed Hoffman, was warned by an FBI agent that he "might get killed" if he revealed what he observed in Dealey Plaza on the day of the assassination.

Three tramps

Some witnesses maintained that three tramps were the assassins.

Mysterious and suspicious deaths took place with witnesses of the Kennedy assassination. Jim Marrs, an American newspaper journalist, later released a list of 103 people he believed had died "convenient deaths" that seemed to be suspicious. Interestingly, the deaths seemed to be grouped around investigations that were handled by the Warren Commission.

On February 23, 1964, a journalist named Dorothy Kilgallen, who was employed by the New York City newspaper Journal American, wrote an article about an interview she had with Jack Ruby at his defense table. She intended to make an additional interview later. She never got the chance. Kilgallen was found dead due to a combination of alcohol and sleeping pills on November 8, 1965, in her home in Manhattan. Bugliosi stated that this was perhaps the most prominent mysterious death cited by assassin researchers.

Thirteen days before the assassination, Joseph Milteer, director of the Dixie Klan of Georgia, was secretly taped telling Miami Police informant William Somerset that a murder of Kennedy was in the working. The threat was ignored by Secret Service personnel when planning the trip to Dallas.

Some assassination researchers state that witness statements indicating a conspiracy were ignored by the Warren Commission. Josiah Thompson declared that the Commission discounted and ignored the accounts given by seven different eyewitnesses who said they saw smoke in the vicinity of the grassy knoll when the assassination was taking place. They also ignored an eighth witness who said he smelled gun powder.

Jim Marrs wrote that the Commission did not seek testimony from eyewitnesses on the triple underpass whose statements pointed to a shooter on the grassy knoll.

The grassy knoll

The Newman family heard gunshots from the grassy knoll during the assassination and heard bullets whistle by them. They were between the grassy knoll and the limousine, so they threw themselves on the ground in an effort to keep from getting shot.

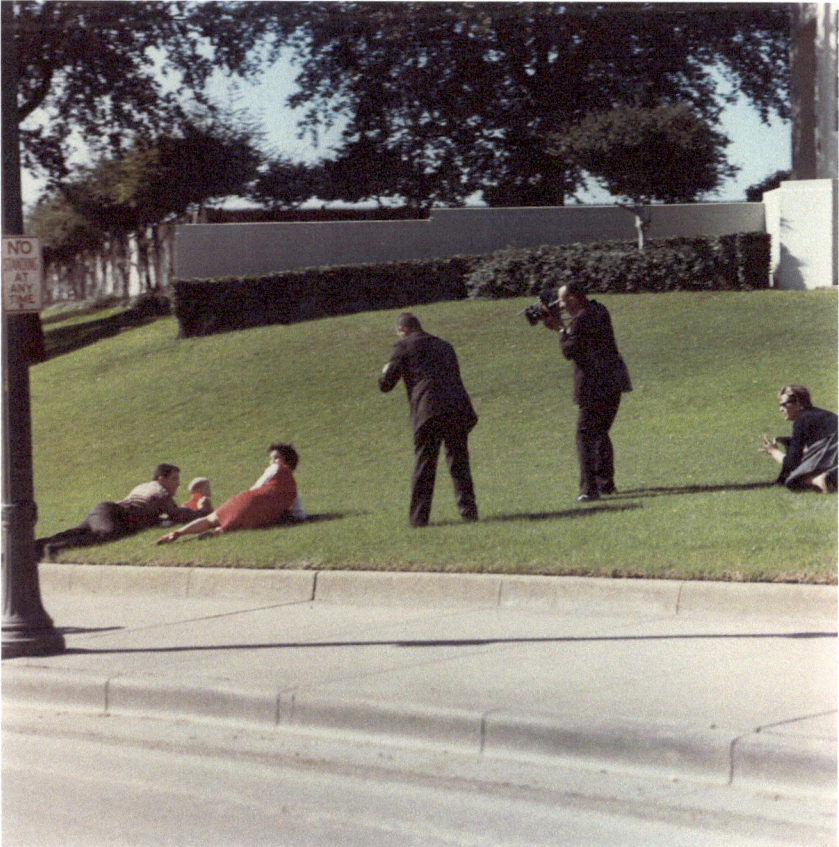

The Newman family on the ground

Dallas Police Officer J.D. Tippit has been named in some conspiracy theories as a rogue CIA agent sent to silence Oswald and to kill John F. Kennedy from the grassy knoll. But Tippit was killed on a local street. The conspiracy theorists believe that Oswald did not kill him. It was made to look as if he did.

The killer hoped that the Dallas police officers would be so angry about the killing that they would shoot Oswald down on sight. They figure that it does not make sense that if Oswald had killed Tippit he would immediately go to a movie theater to watch a movie. The Dallas police would be able to easily corner him there.

The FBI received the bullets used to kill Officer Tippit. They could not match them to Oswald's gun. The spent cartridges matched the gun but not the bullets. The ballistic evidence did not match up.

The Dallas police lifted fingerprints from the passenger side of Tippit's police car. They did not match Oswald's fingerprints.

The main witness to the shooting was Helen Markham. Mark Lane, who was hired to represent Oswald, asked Helen in a recorded conversation to describe Officer Tippit's killer. She stated that he was short and heavy with bushy hair. This description does not match Oswald's appearance.

Helen Markham picked Lee Harvey Oswald out of the police lineup. However, she later stated that she didn't recognize Oswald in the police lineup and that she only picked him out because she had chills run up and down her when she saw him in the lineup. The Warren Commission built most of the case against Oswald around the shooting of Tippit based on her testimony.

Many people, including Richard Buyer, have protested that many documents connected to the assassination have been held back over the years, including reports from inquiries made by the Warren Commission, the House Select Committee on Assassinations, and the Church Committee. Among these documents are the President's autopsy records. Some documents are not slated to be released until 2029. Despite the fact that there was a release of many important documents relating to the assassination in the mid-to-late 1990s by the Assassination Records Review Board (ARRB) under the President John F. Kennedy Assassination Records Collection Act of 1992, some of the documents contain redacted sections. Things like tax return information, which pin pointed various employers and sources of income, has not been released yet.

The long periods of secrecy and the with holding of secret documents even from the Warren Commission has led many people to believe that there was and is a government cover up.

In 2010, journalist Jefferson Morley used the Freedom of Information Act to approach the CIA. The response of the CIA was that they had over 1,100 documents related to the assassination, about a total of 2,000 pages that have not been released due to national security concerns.

Some researchers maintain that physical evidence has been tampered with. This includes various bullets and bullet cartridges and fragments, the presidential limousine windshield, the paper bag in which the Warren Commission said Oswald hid his rifle, the back yard photos showing Oswald holding the rifle, the Zapruder film, the photographs and radiographs obtained at Kennedy's autopsy, and Kennedy's body itself.

T. Jeremy Gun, who work for the Assassination Records Review Board, has said that there are serious problems with the forensic evidence, with the ballistic evidence, and with the autopsy evidence. Gun stated that the Warren Commission believed that Lee Harvey Oswald had committed the crime so that is what they went with. But the big mysteries caused a lot of troubles. The institution which had the best opportunity to get to the bottom of this was the Warren Commission, and they did not do it. Gun said, "Now it's too late to do what should have been done originally."

The suspicions created by government secrecy eroded confidence in the truthfulness of federal agencies in general and damaged their credibility.

16. THE PLOT TO KILL FIDEL CASTRO

There have been weird conspiracy theories about the CIA plotting to assassinate Fidel Castro.

Fidel Castro (right) and Camilo Cienfuegos, 1959
Photo by Luis Korda

A report that was created in 1967 and then declassified, 36 years later proved that these conspiracies were actually true. Throughout the duration of 1960 to 1965 the CIA had attempted to kill Fidel Castro over 600 times. The CIA tried just about everything. Some ideas were plausible ones such as lacing cigars with poison or tiny bombs. Poison was one of their favourite tactics. They tried contaminated air, poisoned pills, poisoned ice cream, and a poison-filled syringe. They even tried to get a woman named Marita Lorenz, Castro's lover, to poison his drink. This plan failed because the poison pills were smuggled inside her container of cold cream and had dissolved before she was able to put them in his drink.

The CIA were so intent on getting at Castro that they even resorted to foolish school boy type pranks. They thought that maybe they could assassinate his reputation as well as his life. So, they thought of things like trying to get him drugged on LSD so that he would make a public fool of himself on the radio. They considered putting thallium salt in his shoes or in his scuba suit in an effort to make all of his hair fall out. They put high bounties on the heads of all the people under him but only a 2 cent bounty on his head. They were hoping that all of these attacks on his reputation might cause him to be so humiliated that perhaps he would commit suicide. Most of these plots the CIA was not that serious about, and they never got past the planning stage.

One far-fetched idea was to to place a particularly beautiful, booby trapped shell in a reef Castro was known to frequent. Castro was a fan of scuba diving, so intelligence agents hoped that Castro would see the shell and think it was so beautiful that he would pick it up. When he did so, it would set off an explosion that would kill him. This idea was so impractical that it was called off.

In the end, the CIA failed at their conspiracy. Castro died in 2016 at the age of 90 due to a heart attack.

17. THE CIA AND AIDS

HIV stands for Human Immunodeficiency Virus. The virus attacks the human immune system. It enters the white blood cells where it reproduces itself. As more and more white blood cells are destroyed and more and more of the virus is produced, the immune system becomes compromised. Eventually the person with the virus is said to have an acquired immunity deficiency syndrome or AIDS.

The AIDS epidemic was first reported in 1981. Since that time, there have been rumors that the AIDS virus was created by the CIA to wipe out homosexuals and African Americans. This conspiracy theory has many high profile believers. South African President Thabo Mbeki once accused the U.S. Government of manufacturing it in military labs. The Nobel Prize winner, Kenyan ecologist, Wangari Maathai, has also accused the U.S. Government of manufacturing it in military laboratories.

Most members of the scientific community believe that the virus jumped from monkeys to humans some time in the 1930s.

18. THE 9/11 COVER UP

On the morning of September 11, 2011, nineteen Al-Qaeda terrorists hijacked four commercial passenger jet airliners, intentionally crashing two of them into the World Trade Center in New York City. The hijackers crashed a third airliner into the Pentagon. The fourth plane crashed in a field near Shanksville, Pennsylvania. Nearly 3,000 victims and the nineteen hijackers died in the attacks.

World Trade Center, New York City - aerial view (March 2001)
Photo by Jeffmock at www.wikipedia.org

World Trade Center, New York City – (September 11, 2001)
Source- Wikimedia Commons

At about the same time the towers were hit, another passenger plane crashed into the pentagon.

World Trade Center, New York City – (September 11, 2001)
Source- https://archive.defense.gov/news Photo by Jim Garamone

Shortly after the event, my daughter, Heather, showed me some verses from the book of Revelation.

Revelation 18:2 "Babylon the great is fallen, is fallen, ….."

Revelation 18:15 "The merchants of these things which were made rich by her shall, stand afar off for the fear of her torment, weeping and wailing."

Revelation 18:16: "and saying, Alas, alas, that great city, that was clothed in fine linen, and purple, and scarlet, and decked with gold, and precious pearls!"

Revelation 18:17: "For in one hour such great riches is come to naught. And every ship master and all the company in ships, and sailors, and as many as trade by sea, stood afar off."

Revelation 18:18: "And cried when they saw the smoke of her burning."

Videos and films of the two planes striking the two towers are famous around the world. The amount of documented evidence has sparked a profusion of conspiracy theories.

Somehow some TV shows were able to show Osama Bin Laden celebrating the event. A later TV show featured Osama Bin Laden and his group of Al-Qaeda and showed that they had planned the whole thing.

Osama bin Laden being interviewed by Hamid Mir, circa March 1997 – May 1998.
Photo by Hamid Mir CC BY-SA 3.0

The leader of the nineteen terrorists was said to be the Egyptian, Mohamed Atta.

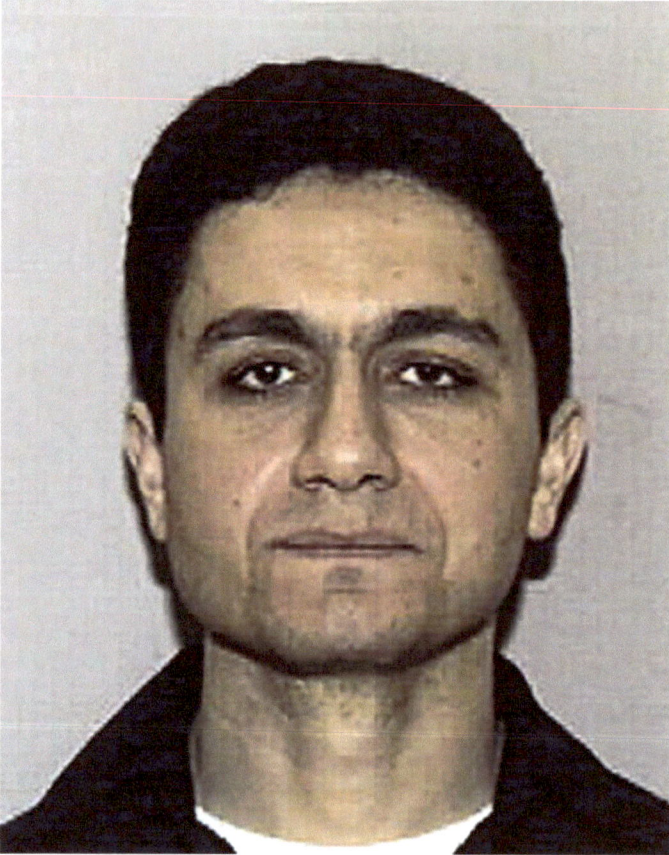

Mohamed Atta al-Sayed, an Egyptian who led the September 11 attacks. Picture from his Florida driver's license. It appeared on the FBI's website shortly after the attacks.

There were many conspiracy theories:

1) It was a false flag operation and the greatest act of treason in human history – perpetrated by the Bush administration as a sort of Pearl Harbor that could be used to justify both the imposition of unprecedented surveillance on the population and wars in Afghanistan and Iraq.

2) The administration knew an attack was planned and let it happen.

3) It was an Israeli plot carried out by Mossad.

4) The towers collapsed because of "controlled demolitions" perpetrated by a person or persons unknown but probably by the CIA.

5) The motive was to profit from "insider trading" and manipulation of the stock values of United Airlines and American Airlines.

People had good cause to be suspicious. The great escalation of the Vietnam War was justified by something that never happened; the fake news of the "Gulf of Tonkin incident". The invasion of Iraq was justified by non-existent "weapons of mass destruction".

The official inquiry involved a vast cover-up that remains in place, indicated by the redacted pages in the 9/11 Commission Report about the U.S.A.'s close friends in Saudi Arabia. Not only was bin Laden a Saudi Arabian, and a member of a vastly wealthy dynasty described as the "Rockefellers", but fifteen of the 9/11 terrorists, most of the team members, were Saudi nationals. This was hushed up and blame for the attack deliberately misdirected to Iraq.

High ranking Saudis were secretly smuggled out of the U. S. while all other international flights were grounded.

An important contention by the theorists was that the burning fuel from the airplanes which had crashed into the towers would not burn hot enough to melt the steel girders and supports so that the buildings would crash. Therefore, the theorists maintained that the two towers had been rigged with explosives beforehand. However, there wasn't any evidence of planted explosives.

Although the fuel did not burn hot enough to bring down the towers, it started fires all through the buildings which burned a lot hotter than the jet fuel. This was still not hot enough to melt the steel supports of the towers. It was, however, hot enough to destroy the integrity of the steel. The steel did not melt but its loss of integrity caused it to bend. When it had bent far enough, it snapped. The steel broke into pieces causing the twin towers to collapse.

19. PAUL MCCARTNEY IMPOSTER

Conspiracy minded Beatlemaniacs say, Paul McCartney secretly died in 1966, in a fatal car accident. The remaining members of the Beatles covered it up so the show could go on. They hired a replacement.

Press photo of the Beatles in 1967 (Paul is on the right)
Source - Wikimedia Commons

The conspiracy theorists say he never wrote "Maybe I'm Amazed". They say he never clashed with Yoko, became a vegetarian, and never fathered any of his children. When Queen Elizabeth knighted him in 1997, she was actually knighting someone else. The Beatles had hired someone else named Bill Shears, who looked like Paul, sang like him, and had his personality.

If Paul was really dead, his impostor was still very active. He met and married Linda Eastman, and they had four children together. Linda died of breast cancer in 1998. He released a live album in 1993 called "Paul is Live" and produced more than 20 solo albums. That is not counting the ones released by Wings.

He endured a horrible divorce from Heather Mills. This may have made him wish he were really dead or maybe that he was still Bill Shears. So, who is the real McCartney? Will the world ever know?

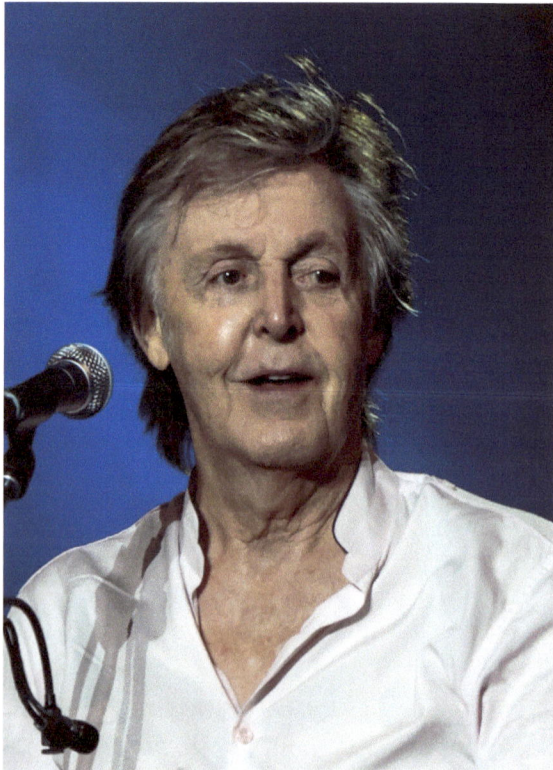

Paul McCartney in October, 2018
Source - Wikimedia Commons - Author Raph_PH

20. JESUS AND MARY MAGDALENE

In 2003, the book, "The Da Vinci Code" by Dan Brown appeared in book stores. In this fiction story, it alludes that Jesus was married to Mary Magdalene, and that they had a child together. The story also forwards the notion that the church had covered it up.

The established churches consider the marriage of Jesus and Mary Magdalene to be a conspiracy theory. The church authority does not like that they were married. Whether there is a blood line going back to Jesus and Mary is something that has been debated for a long time.

In 2012, a Harvard University scholar and historian of early Christianity, Karen L. King went to a conference in Rome and presented writings on a small piece of a papyrus. The piece of papyrus was dated as being 1,300 years old. On it, in the ancient Coptic language, were the words, ' Jesus said to them, "My wife."   The fragment came from an anonymous collector.

In 325 AD, there was also a conference chaired by the Emperor of Rome, Constantine. It was a council of Bishops in a city in Nicea. The Bishops came from all over the Roman Empire. The aim of this meeting was to obtain a consensus in the church of various issues. The decision as to which gospels were to be included in the Bible was decided at that time. All other gospels were to be destroyed. The church ordered the burning of all the other gospels.

In Upper Egypt, someone, most likely a monk from the monastery of St. Pachomius, took the offending books and buried them in a jar. This acted as a time capsule. In December 1945, a peasant found a clay jar with some of the scrolls in it, and in 1946, more of the Dead Sea Scrolls were found in a cave.

There are many different controversies between the Gnostic gospels and the gospels presented in the Orthodox Church. The Orthodox Church degrades women. It presents Mary Magdalene as a prostitute and a sinner, which is a complete lie. The Gnostic gospels present her as an Apostle and leader in the early church.

Upon researching the gospels and the culture and customs of that time, the proof of the historical marriage between Jesus of Nazareth and Mary Magdalene becomes overwhelming.

None of the four gospels called Jesus celibate. These gospels call Jesus Rabbi. Both then and now a Rabbi is married and has children.

After the crucifiction of Jesus, it was Mary Magdalene who went to wash and anoint His body. At that time a woman would not wash the naked body of a dead man unless she were a family member or a wife.

A compelling argument in the debate of whether Jesus married Mary Magdalene and had children together, lies in an archeological find in 1980, in Talpiot, just outside of Jerusalem. Inside a burial tomb, ten limestone coffins were discovered. Six of these coffins were inscribed with names. One was inscribed with the Hebrew/Aramaic name "Jesus son of Joseph". A second one was inscribed with the name "Mariamene," which is a Greek variant of "Mary" associated in all of Greek literature with only one woman — Mary the Magdalene. Surprisingly, one of the six was that of a child and had the name "Judah, son of Jesus" carved on it.

With the Gnostic gospels, there is now written evidence that Jesus married Mary Magdalene and that they had children together. The established church authorities still call this a conspiracy theory.

21. HOLOCAUST REVISIONISM

Revisionists state that nearly 6 million Jews were not killed by the Nazis during the Holocaust. One of these revisionists, Iranian President Mahmoud Ahmadinejad, has called the Holocaust a "myth". He wants Germany and other European countries to provide land for a Jewish state instead of Palestine.

Most revisionists do not disagree that the Jews were placed in internment camps. They maintain that the number of deaths were greatly exaggerated. They state that gas chambers were only a rumour or if they existed they were not powerful enough to kill. They deny evidence and history. The revisionists claim that typhus was raging at the time and that all those photographs of skinny people and corpses stacked like cord wood were actually of Czechs and Poles and Germans who died of typhus.

Bodies in Buchenwald Concentration Camp

Holocaust denial is considered to be a serious societal problem in many places where it occurs. It is illegal in several European countries and in Israel. Holocaust denial is sponsored by some middle Eastern governments including Iran and Syria.

Most Holocaust deniers claim that the Holocaust is a hoax or an exaggeration perpetuated by the Jews designed to advance their own cause at the expense of other people. For this reason Holocaust denial is considered to be an antisemitic conspiracy theory.

22. THE SANDY HOOK ELEMENTARY SCHOOL MASSACRE

On December 14, 2012, twenty year old Adam Lanza killed his mother, fifty-two year old Nancy Lanza, in her bed by shooting her four times in her head.

Adam then drove to the elementary school in his mother's car. The doors of the school were normally locked every day at 9:30 a.m. Using his mother's Bushmaster XM115 E2S rifle and ten magazines with 30 rounds each, Adam shot his way through a glass panel next to the locked front entrance doors at 9:35 a.m. Then, he shot and killed 20 first graders; 8 boys and 12 girls, as well as 6 teachers. All of the victims, except two, were shot multiple times.

At 9:40 a.m. Adam Lanza committed suicide by shooting himself in the head.

Police arrive at Sandy Hook Elementary, after the shooting on December 14, 2012

The Newton Police arrived at the school at 9:39 a.m. approximately four and a half minutes after the 911 call. The Connecticut State Police arrived at the school at 9:46 a.m. The police first entered the school at 9:45 a.m. approximately 14 minutes after the shooting started.

Conspiracy theorists have claimed that the shootings never happened. James Fetzer and Mike Palazek wrote a book called "Nobody Died At Sandy Hook", claiming that the shooting never happened. Lenny Pozner, whose six year old son died in the shooting, filed a claim against the publisher. Dave Gahary, the principal officer at publisher Moon Rock Books, has said, "My face-to-face interactions with Mr. Pozner has led me to believe that Mr. Pozner is telling the truth about the death of his son. I extend my most heartfelt and sincere apology to the Pozner family".

A judge in Wisconsin issued a summary judgment against the authors James Fetzer and Mike Palazek.

Lenny Pozner has been fighting back for years against hoax believers who have harassed him. They have claimed he was an actor and that his son never existed. He has also received death threats. He spent years getting Facebook and others to remove conspiracy videos and set up a website to debunk conspiracy theories. In 2019, a U.S. jury awarded $450,000.00 to Lenny Pozner in a defamation lawsuit against James Fetzer and Mike Palazek.

Other people who also lost relatives in the Sandy Hook massacre on December 14, 2012, have also been harassed. Robbie Parker, whose 6 year old daughter Emilie was among the 20 first graders killed, spent years ignoring people who called him a crisis actor. His family moved to the West Coast but the harassment continued. He received letters from people who found his address. He was once stopped in a parking garage by a man who berated him and told him the shooting never happened.

Robbie Parker became tired of the harassment and began fighting back. He is now in a lawsuit against conspiracy theorist Alex Jones. He has testified before Congress and pushed for changes on social media platforms such as YouTube which announced that it will prohibit videos that deny the Sandy Hook shooting and other "well-documented events".

The Sandy Hook Elementary School makeshift memorial on Washington Avenue in Sandy Hook, Conn., 12 days after the shootings at Sandy Hook Elementary School.
Source- wikipedia.org Photo by Bbjeter

Pozner is the lead plaintiff in lawsuits of at least nine cases filed against Sandy Hook deniers in federal and state courts in Connecticut, Florida, Texas, and Wisconsin.

In the case against Alex Jones, the families of eight victims and a first-responder say they have been subjected to harassment and death threats from the followers of Jones. A Florida woman, Lucy Richards, was sentenced to five months in jail for sending Pozner death threats. She was also banned from visiting web sites run by conspiracy theorists, including Fezner.

23. SOME FINAL THOUGHTS

The mass media can significantly affect our understanding of what is happening in the world. Some award-winning journalists, in trying to report major news items, have had these stories shut down by corporate media ownership. These stories have included "dangerous hormones in our milk", "the shooting down of TWA flight 800 by the military", "thousands of prisoners of war abandoned in Vietnam", "the CIA involvement in the narcotics trade", and more.

The internet has destroyed the traditional media gate keepers and gave everyone a chance to indulge in news that confirmed their own preexisting ideas. How is the public now sorting fact from fiction?

In the end, many of the conspiracy theories come from people who simply do not understand the principles of photography, physics or other sciences. We must take into consideration that there is a difference between a skeptic and a denier. A skeptic is one who insists on evidence, and accepts it when it is given even if it doesn't support what they originally thought to be true. Deniers blindly and stubbornly stick to their own beliefs either without any evidence or refuse to accept valid evidence when it is presented to them.

Mysterious and suspicious deaths took place with witnesses of the Kennedy assassination. Jim Marrs later presented a list of 103 people he believed had died "convenient deaths" under suspicious circumstances. He noted that the deaths were grouped around investigations conducted by the Warren Commission.

T. Jeremy Gun, who investigated the John F. Kennedy assassination, has said that there are serious problems with the forensic evidence, with the ballistic evidence, and with the autopsy evidence. Gun stated that the Warren Commission believed that Lee Harvey Oswald had committed the crime so that is what they went with. But the big mysteries caused a lot of troubles. The institution which had the best opportunity to get to the bottom of this was the Warren Commission, and they did not do it. Gun said, "Now it's too late to do what should have been done originally."

Suspicions created by government secrecy has eroded confidence in the truthfulness of federal agencies in general and damaged their credibility.

Elderly Holocaust survivors have had to face some Holocaust deniers.

The official inquiry of the 9/11 attack on September 11, 2001, involved a vast cover-up that remains in place, indicated by the redacted pages in the 9/11 Commission Report about the U.S.A.'s close friends in Saudi Arabia.

People who lost relatives in the Sandy Hook massacre on December 14, 2012, have been harassed by conspiracy theorists denying that the event ever took place. Some have even received death threats over it.

The so-called conspiracy theories are destructive to democracy. The author, Nancy Rosenblum, has stated that conspiracism can destroy democracy on its own. That it does not take an alternative political ideology like communism, or authoritarianism, or fascism to destroy democracy. Conspiracism can do it all on it's own.

REFERENCES

Steuart from *South Africa The Good News* (July 13, 2019) "How Much Do You Know About Madiba's Life?"
Retreived from
https://www.sagoodnews.co.za/much-know-madibas-life/

The Mandela Foundation " Biography of Nelson Mandela"
Accessed on January 31, 2020
https://www.nelsonmandela.org/content/page/biography

Caitlin Aamodt (February 16, 2017) "Collective False Memories: What's Behind the 'Mandela Effect'?"
Retreived from
https://www.discovermagazine.com/mind/collective-false-memories-whats-behind-the-mandela-effect

R. E. Cox, and A. Barnier (July 17, 2015) "A Hypnotic Analogue of Clinical Confabulation"
Retreived from
https://nvvh.com/wp-content/uploads/2015/07/IJCEH-63-3-01.pdf

Gill & Brenman, (1959); E. R. Hilgard, (1977) "Multiple Predictors of Hypnotic Susceptibility"
Retreived from
https://psycnet.apa.org/doiLanding?doi=10.1037%2F0022-3514.53.5.948

Rachel Nall, RN, MSN, CRNA (March 13, 2020) "The Mandela Effect: How False Memories Occur"
Retreived from
https://www.healthline.com/health/mental-health/mandela-effect

Blake Bakkila (August 6, 2019) "40 Mandela Effect Examples"
Retreived from
https://www.goodhousekeeping.com/life/entertainment/g28438966/mandela-effect-examples/

Patrick J. Kiger "10 Reasons The Multiverse Is A Real Possibility"
Retreived from
https://science.howstuffworks.com/10-reasons-multiverse-is-real-possibility.htm

Elizabeth Howell (May 10, 2018) "Parallel Universes: Theories & Evidence"
Retreived from
https://www.space.com/32728-parallel-universes.html

Christopher Hudspeth (November 15, 2016) "9 Strange Stories That Might Make you Believe In Parallel Universes"
Retreived from
https://www.buzzfeed.com/christopherhudspeth/9-eerie-stories-of-people-who-mayve-experienced-a-parallel-d

Becky Little (July 18, 2019) "The Wildest Moon Landing Conspiracy Theories Debunked"
Retreived from
https://www.history.com/news/moon-landing-fake-conspiracy-theories

Chris Bell (February 1, 2018) "The people who think 9/11 may have been an 'inside job'"
Retreived from
https://www.bbc.com/news/blogs-trending-42195513

Brooke Bowman (January 30, 2012) "Shipwreck Hunters Stumble Across Mysterious Find"
Retreived from
https://www.cnn.com/2012/01/28/world/europe/swedish-shipwreck-hunters/index.html

Richard Buyer (July 15, 2009) "Why the JFK Assassination Still Matters"
Published by Wheatmark
ISBN-13: 978-1604941937

Duncan Campbell (November 26, 2016) "Close But No Cigar: How America Failed To Kill Fidel Castro"
Retreived from
https://www.theguardian.com/world/2016/nov/26/fidel-castro-cia-cigar-assasination-attempts

Theresa Waldrop (January 27, 2020) "Sandy Hook Denier Charged With Illegally Possessing ID Of Victim's Dad"
Retreived from
https://www.cnn.com/2020/01/27/us/sandy-hook-denier-wolfgang-halbig-arrest/index.html

Joyce Chen (November 6, 2017) "Paul McCartney is Dead: Music's Most WTF Conspiracy Theories Explained"
Retreived from
https://www.rollingstone.com/music/music-news/paul-mccartney-is-dead-musics-most-wtf-conspiracy-theories-explained-120340/

John Kerrison (May 23, 2014) "Everyday Chemistry: The Story Behind The Greatest Beatles' Albums That Never Existed"
Retreived from
https://medium.com/much-stranger-than-fiction/everyday-chemistry-the-story-behind-the-greatest-beatles-albums-that-never-existed-517fb5f415fd

Elaine H. Pagels (September 19, 1989) "The Gnostic Gospels"
Published by Vintage; Reissue edition
ISBN-13: 978-0679724537

Karen King (November 1, 2003) "The Gospel Of Mary Of Magdela"
Published by Polebridge Pr Westar Inst
ISBN-13: 978-0944344583

Jacob Heller (January 2015) "Rumors and Realities: Making Sense of HIV/AIDS Conspiracy Narratives And Contemporary Legends"
Retreived from
https://www.ncbi.nlm.nih.gov/pmc/articles/PMC42659 31/

United States Holocaust Memorial Museum, Washington, DC "Holocaust Deniers and Public Disinformation"
Accessed on January 29, 2020 from
https://encyclopedia.ushmm.org/content/en/article/ho locaust-deniers-and-public-misinformation

Jefferson Morely (April 30, 2019) "JFK Records Suit Tests CIA Secrecy On Assassination"
Retreived from
https://www.justsecurity.org/63826/jfk-records-suit-tests-cia-secrecy-on-assassination/

Various contributers "List of Reported UFO Sightings"
Accessed on June 14, 2020 from
https://en.wikipedia.org/wiki/List_of_reported_UFO_s ightings

Julia   Shaw, (September 13, 2016).   "The Memory
Illusion: Why You Might Not Be Who You Think You
Are"
Published by Doubleday Canada
ISBN-13: 978-0385685290

Bill Sloan (September 6, 2012)"The Kennedy
Conspiracy: 12 Startling Revelations About the JFK
Assassination" Kindle Edition
ASIN: B0097CSV2Q

Jackie Mansky (March 7, 2016) "Eight Secret Societies
You Might Not Know"
Retreived from
https://www.smithsonianmag.com/history/secret-
societies-you-might-not-know-180958294/

Time (2019) "Separating Fact From Fiction: The JFK
Assassination"
Retreived from
http://content.time.com/time/specials/packages/article
/0,28804,1860871_1860876_1861003,00.html

Max Tegmark for Scientific American (May 2003)
"Parallel Universes: Not Just A Staple Of Science
Fiction, Other Universes Are A Direct Implication Of
Cosmological Observations"
Retreived from
https://space.mit.edu/home/tegmark/PDF/multiverse_
sciam.pdf

CERN Accelerating Science "The Large Hadron Collider"
Accessed on February 24, 2020
https://home.cern/science/accelerators/large-hadron-collider

Dennis Overbye of The New York Times (December 15, 2015) "Physicists in Europe Find Tantalizing Hints of a Mysterious New Particle"
Retreived from
https://www.nytimes.com/2015/12/16/science/physicists-in-europe-find-tantalizing-hints-of-a-mysterious-new-particle.html

Sylvia Strojek (July 10, 2018) "Here are the Canadian cities with the most UFO sightings in 2017: survey"
Retreived from
https://globalnews.ca/news/4323986/ufo-sightings-canada-2017-ufology/

Staff of The canadian Press (March 12, 2014) "About 1,100 UFOs spotted in Canada last year, survey says"

Retreived from
https://globalnews.ca/news/1202634/about-1100-ufos-spotted-last-year-survey/

The Shag Harbour Incident Society "The Incident"
Accessed on February 16, 2019
https://shagharbourincident.wordpress.com/the-incident/

Suzette Belliveau CTV News Atlantic Reporter (October 5. 2019) "Shag Harbour Incident continues to fascinate small N.S. fishing village"
Retreived from
https://atlantic.ctvnews.ca/shag-harbour-incident-continues-to-fascinate-small-n-s-fishing-village-1.4625623

Gerald Chung (May 23, 2016) "James Earl Jones recalls "Luke, I am your father."
Retreived from
https://www.youtube.com/watch?v=GQ1mmkKb_BQ

Gerald Chung (October 4, 2016) "Hallmark Ornament plays 'Luke, I am your father.'"
Retreived from
https://www.youtube.com/watch?v=dEnRV9P74XE

ABOUT THE AUTHOR

Carsten Jorgensen dedicated thirty years in studying and managing fisheries for the Ontario Ministry of Natural Resources.

Upon graduation from Queen's University in Kingston, Ontario in 1966, he accepted a biologist position on Lake Temagami with the Ontario Department of Lands and Forests.

In 1968 he also started work on Lake Nipissing. In 1970, Mr. Jorgensen was working full time as the Lake Nipissing Fisheries Assessment Unit Biologist.

In 1970 he married Brenda Black, daughter of Ontario Conservation Officer, Gordon Black.

In 1996 he retired and now enjoys spending his time playing chess, playing darts, watching Star Trek and documentaries, and writing books.

OTHER TITLES BY CARSTEN R. JORGENSEN

If you enjoyed this book by Carsten R. Jorgensen, you may also enjoy these other books that he has written:

The Saga Kings -
ISBN-13: 978-09949338-0-5

Trying To Work For The M.N.R. -
ISBN-13: 978-0-9949338-1-2

My World War Two Adventures In Denmark -
ISBN-13: 978-0-9949338-2-9

One School, Two School, Old School, New School -
ISBN-13: 978-0-9949338-3-6

Spiritual Encounters And Other Strange Stories From The Little Red School-House -
ISBN-13: 978-0-9938776-3-6

Fishes Of Lake Nipissing -
ISBN-13: 978-0-9949338-4-3

Dragons Of The World And Where They Roam -
ISBN-13: 978-0-9949338-5-0

Myths, Mythology, and Faith -
ISBN-13: 978-0-9949338-6-7

Recent Extinctions -
ISBN-13: 978-0-9949338-7-4

Communications In Nature -
ISBN-13: 978-0-9949338-8-1

Or check Good Reads for any of his upcoming books:
www.goodreads.com/author/show/14680643.Carsten_R_Jorgensen

www.ingramcontent.com/pod-product-compliance
Lightning Source LLC
Chambersburg PA
CBHW041308210326
41599CB00003B/28